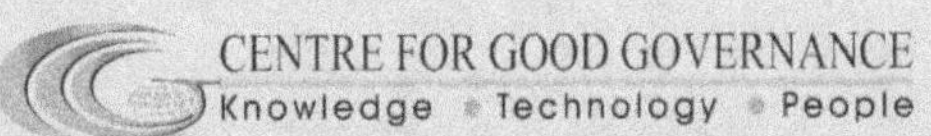

Capacity Building for

Climate Smart Agriculture

Edited by

Devi Prasad Juvvadi

BS Publications

Capacity Building for Climate Smart Agriculture *by J. Devi Prasad*

© 2016

Disclaimer

The Publishers and Editor do not claim any responsibility for the accuracy of the data, statements made, and opinions expressed by authors. The authors are solely responsible for the contents published in the paper.

Published by BS Publications in association with International Food Policy Research Institute, New Delhi, *"Centre for Good Governance, Hyderabad"*

Published by

CENTRE FOR GOOD GOVERNANCE
Knowledge • Technology • People

(Dr. MCR HRD Institute, Telangana)
Road No. 25, Jubilee Hills,
Hyderabad 500033
Telangana (India)
www.cgg.gov.in
Phone: +91 40 2354 1907 / 09
Fax: +91 40 2354 1953
Email: info@cgg.gov.in

INTERNATIONAL
FOOD POLICY
RESEARCH
INSTITUTE

NASC Complex, CG Block
Dev Prakash Shastri Road
(Opp. Todapur), Pusa,
New Delhi 110012 India
Phone: +91 11 25846566
Email: ifpri-NewDelhi@cgiar.org

BSP BS Publications

A unit of **BSP Books Pvt. Ltd.**

4-4-309/316, Giriraj Lane, Sultan Bazar,
Hyderabad - 500 095
Phone : 040 - 23445605, 23445688
e-mail : info@bspbooks.net
www.bspbooks.net

Citation : Devi Prasad Juvvadi. (2016). Capacity Building for Climate Smart Agriculture Published by BS Publications.

ISBN : 9789352301003

Dedicated to
the vision of   **K. Chandrashekar Rao**
in achieving statehood
and for his commitment
to transform the lives of farmers
of Telangana state

Contents

Message by K. Ramakrishna Rao...(vii)

Agriculture is Part of the Solution to Climate Change by P.K. Joshi.............(ix)

Transform Agriculture for Climate Change by P.K. Aggarwal.....................(xi)

Insight to Climate Smart Agriculture by Kirit N Shelat.............................(xiii)

Farming need's "Climate-smart" Revolution by J.P. Mishra.....................(xv)

Foreword...(xvii)

Preface...(xix)

Acknowledgement...(xxiii)

1. **Capacity Building in Extension: Key to Climate Smart Agriculture**
   Devi Prasad Juvvadi... 1

2. **Basic Concepts of Climate Change with Global and Local Perspectives**
   Ch. Srinivasa Rao and Arun K. Shankar..15

3. **Impact of Climate Change on Agriculture and Food Security**
   A.V.R. Kesava Rao and Suhas P Wani...27

4. **Vulnerability of Indian Agriculture to Climate Change and its Assessment**
   V. K. Sehgal, H. Pathak and S. D. Singh..53

5. **Need for Building Capacities of Institutions and Individuals to Deal with Climate Change**
   G. V. Ramanjaneyulu..67

6. **Impact of Climate Change on Soil Properties and their Mitigation Strategies**
   Pragati Pramanik..87

7. **Climate Change and Drought Governance in Dry Land Agriculture**
   Devi Prasad Juvvadi..97

8. **Climate Smart Agriculture for Building Resilience and Improving Livelihoods in Rainfed Areas**
   Suhas P Wani...129

9. "Climate Resilient Agricultural Practices – NICRA Experiences"
   Y.G. Prasad 133

10. Soil Management for Climate-Resilient Agriculture
   H. Pathak 145

11. Crop Specific Technologies for Climate Resilient Agriculture
   J. V. N. S. Prasad 163

12. Farmer's Knowledge, Attitude, Sources of Information and Adaption Measures towards Climate Change
   K. Ravi Shankar, K. Nagasree, G. Nirmala, M.S. Prasad and
   Ch. Srinivasa Rao 181

13. Community Based Approaches to Climate Smart Agriculture
   Rajeswar Jonnalagadda 197

14. Financial Instruments Available in Public and Private Sectors for Climate Smart Agriculture
   R. Sudhakar Rao 209

15. Risk Transfer Mechanism in Climate Smart Agriculture
   C. Kiran Kumar 229

16. Planning and Preparation of Projects for Climate Smart Agriculture under NMSA
   K. Padmaja 235

Appendix – I: National Mission for Sustainable Agriculture (NMSA) Mission Implementation Plan (MIP) 249

Group Activity for Participants: Problem Tree Analysis – Know the World around You 251

Glossary of Common Climate Change Terms 257

About Contributing Authors 277

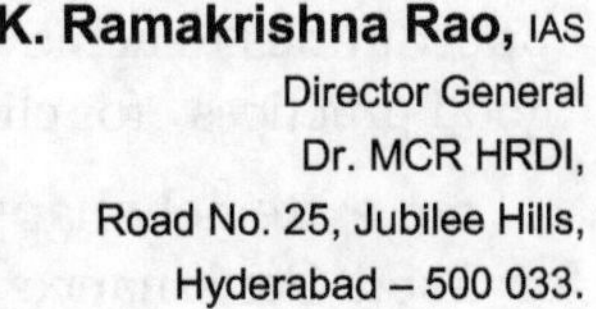

K. Ramakrishna Rao, IAS
Director General
Dr. MCR HRDI,
Road No. 25, Jubilee Hills,
Hyderabad – 500 033.

Message

Climate change is a reality and the vulnerability of Indian agriculture to climate change is well we acknowledged. The need for a climate resilient approach is critical for India where more than 80 percent agriculturists are small holder farmers who have lesser access to knowledge. Climate change presents a daunting task for the policy makers, researchers, extension staff and farmers as to how to mitigate this risk. More importantly, increase in food production and farm profitability must take place in a sustainable manner while combating this intimidating deviant environment.

Knowledge about the technologies and processes that we choose to mitigate climate change are very important and currently gap exists about Climate Smart Agriculture (CSA) practices. Government officials in agriculture departments need to enhance their capacities and bridge the knowledge gap for achieving sustainable agriculture productivity and food security in the fast changing scenario.

The Centre for Good Governance in association with International Food Policy Research Institute (IFPRI) is organizing a three day capacity building programme on "Climate Smart Agriculture" for mid and senior level agricultural extension staff of different states. The program seeks to integrate knowledge including research and action for and strengthen the technical and policy capacities of extension functionaries to help farmers at the farming level and as well as framing policies. It helps to respond to the need to deepen the understanding of the link between climate change and the agriculture and to look at intervention opportunities to achieve adaptation and mitigation and proliferate "good practices" for climate smart agriculture.

I am extremely happy to inform that as part of training, Centre for Good Governance has developed a capacity building manual on "Climate Smart Agriculture" covering essential and crucial issues related to climate change adaptation for agriculture.

The book is expected to enlighten the participants and other readers with climate change and its impact and how CSA can reduce the climate change risk. I wish the programme a grand success and hope that the knowledge provided during this programme will reach the farmers and other extension staff.

Date: 23rd July, 2015

K. Ramakrishna Rao, IAS
Director General, CGG, Hyderabad

Dr. P. K. Joshi
Director (South Asia)
NASC Complex, CG Block
Dev Prakash Shastri Road
(Opp. Todapur), Pusa,
New Delhi 110012 India

Agriculture is Part of the Solution to Climate Change

Agriculture is highly vulnerable to even short term weather change, therefore even a small shift in climate poses direct threat to farmers, who not only have to secure their livelihood but also need to produce sufficient food to feed the growing population. Despite significant achievement in food grain production, farmers today face the challenge of deteriorating land, water and soil, along with growing impact of climate change thus deepening the complexity for their sustainability.

With major presence of small holder farmers in India, who are vulnerable to the changing climate shocks, need timely support to ensure their food security and they also need to be prepared for uncertain future. Efforts are needed to develop interventions on Climate Smart Agriculture (CSA) to counter climate change risks. CSA brings in together improved technologies, value added advisory services, application of information and communication technology and agricultural insurance, which improves the adaption capacity of farmers against climate change and also minimize the greenhouse gas emission. Overall CSA (i) raises agricultural productivity and farm income, (ii) minimizes risk that arise due to climate change and (iii) reduces green house gas emissions.

Signals of the shifting climate change are visible across the country, but the extent of damage differs, therefore government policies, new technologies and adaptive measures should blend together to reach farmers. Process such as, training of trainers to develop capacities of progressive farmers, NGOs working with farmers and the extension workers on Climate Smart Agriculture (CSA) is a step forward.

International Food Policy Research Institute (IFPRI) as part of this on-going research work with Climate Change, Agriculture and Food Security (CCAFS) and Centre for Good Governance (CGG) has organised capacity building in Hyderabad. The three days training workshop covered four modules. (i) orientation about climate change and its impact, (ii) the need for CSA as resilient mechanism, (iii) financing opportunities in CSA, and (iv) project development under National Mission on Sustainable Agriculture. Trainees under this program were will be encouraged to acquire the need based knowledge on CSA to impart training to a large number of farmers in their communities.

I hope that the training workshop has helped in filling the existing knowledge gaps to solve some of the unanswered questions. This publication provides essential information on impacts of Climate Change and how to make agriculture and allied sectors part of the solution to negative impacts of the climate change.

Dr. P K Joshi
Director South Asia,
International Food Policy Research Institute, New Delhi

Prof. Pramod K. Aggarwal
Regional Program Leader (South Asia)
International Water Management Institute
New Delhi Office; NASC Complex; CG Block
Dev Prakash Shastri Marg, Pusa; New Delhi

Transform Agriculture for Climate Change

Several recent reports including those from IPCC indicate a probability of 10-40% loss in crop production in India and other countries of South Asia with increases in temperature by 2070-2100 and decrease in irrigation water unless steps are started now to increase our adaptive capacity. There could be significant losses in some crops such as wheat even in short-term. Droughts, floods, tropical cyclones, heavy precipitation events, hot extremes, and heat waves are known to negatively impact agricultural production, and farmers' livelihood. The projected increase in these events will result in greater instability in food production and threaten livelihood security of farmers. Increased production variability could be perhaps the most significant impact of global impact change. Early signs of decrease in yields due to changing weather have started becoming visible.

Increasing production to meet our rising food demand in a rapidly changing scenario of demand and supplies, availability of land, water and other inputs for agriculture in a scenario of climate change become very crucial. Several technological, institutional and policy interventions have been proposed that can help us adapt to climate change as well as to current and future weather variability. These interventions are often complex and require capacity strengthening of different stakeholders including farmers, extension workers, NGOs and even policy advisors to understand the climate change, its association with current climatic risks, and the options available to adapt in specific circumstances. In this context efforts of Centre for Good Governance (CGG) and the International Food Policy Research Institute (IFPRI) in documenting various options and strengthening capacity of middle and senior level Extension Officials of Agriculture Departments of Karnataka, A.P. and Telangana is a very timely and relevant initiative. Such capacity building will help them make more informed and responsible decisions which eventually will lead to better preparedness of farmers to ensure their livelihoods while simultaneously addressing national food security goals. The document also highlights that cooperation among various agencies will be needed to make a transition towards climate-smart agriculture through national approaches such as National Mission for Sustainable Agriculture for transforming agriculture.

The new book aims at strengthening knowledge base of agriculture extension functionaries, building institutional capacity, leveraging financial resources both state and central and developing strategies for climate smart agriculture.

Prof. Pramod K. Aggarwal
New Delhi

Dr. Kirit N Shelat, IAS (Rtd.)
Executive Chairman (NCCID)
Patel Block, Rajdeep Electronic's Compound,
Near Stadium Six Road, Navrangpura, Ahmedabad-
380 014

Insight to Climate Smart Agriculture

Climate Change poses the greatest challenge to sustainable agriculture and food security in the coming decades. There is need for synergy between technology and public policy for overcoming successfully the adverse impact of changes in temperature, precipitation and sea level on crop and animal husbandry, fisheries and forestry.

One major realization has been that Indian farmers have immense capacity to adapt and accept the climate change. India's development administration, agriculture scientists and civil society members have worked hand in hand with them to face adverse situation that existed at the time of Independence. The Country's economy has been transformed and has been able to develop sustainable agriculture with increased growth rate. This has enabled us to meet food scarcity problem. We are aware that country depended on imports of wheat in initial years and today we are the exporter of wheat.

In the present era, climate change has brought out new and difficult challenges. They need careful understanding of its impact and way-out.

Farmers need to understand them and adapt to new strategy. This is difficult but not insurmountable task. However, this will need considerable efforts on the part of all stakeholders. Challenges are of diverse nature and are un-predictable but with the experience of our farmers and expertise of our scientists, it is possible to make agriculture smarter. There is a growing knowledge about methods of promoting Climate Smart Agriculture and this book provides useful insights into the various steps we should take to insulate farming and farmers from the impact of Climate Change.

Dr. Kirit N Shelat (IAS Rtd.)
Ahmeddabad

21st October, 2014

भारत सरकार
नीति आयोग, संसद मार्ग
नई दिल्ली–110 001
Government of India
NATIONAL INSTITUTION FOR TRANSFORMING INDIA
NITI Aayog, Parliament Street
New Delhi-110 001

Dr. J.P. Mishra
Adviser (Agriculture)
NITI Aayog, New Delhi
Government of India

Farming needs "Climate Smart" Revolution

Climate change is the most challenging global phenomenon adversely impacting the agriculture today. It has widespread impacts both spatially and temporally in addition to the prevalent usual abiotic stresses. Climate smart agriculture becomes a necessity to combat the increased abiotic stresses and challenges for achieving the sustainable development goals. Farming needs "climate smart" revolution with location specific approaches that are important to ensure sustainable agricultural production for food and nutritional security and minimise the inflation of food prices. Understanding the Climate smart agriculture is highly required to bring appreciation amongst policy makers, administrators and academicians to make effective and efficient progress towards food security and sustainable agriculture.

Climate smart agriculture has different connotations at different levels and is concerned with the potential to increase sustainable production of food grains and enhancing the resilience of farming systems to adverse impacts while at the same time help in the mitigation of climate change through reduction in greenhouse gas emission reductions and methods of carbon sequestration. At the local level it could mean providing better opportunities for higher production through improved management techniques. At national level it could be orienting towards the provision of a framework that incentivizes sustainable management practices. Globally, it may relate to setting up of rules in relation to governance of climate change. The climate smart agriculture through adaptation offers opportunities for small and marginal farmers, for greater food security and increased income along with greater resilience.

The publication is a cogent, lucid and well structured compilation of material related to climate smart agriculture. I am happy to learn that Centre for Good Governance (CGG), Hyderabad and International Food Policy Research Institute (IFPRI), New Delhi has brought out this publication on Climate Smart Agriculture. I congratulate the leadership and the vision of the International Food Policy Research Institute (IFPRI), New Delhi and Centre for Good Governance (CGG), Hyderabad for bringing out this timely publication and hope that it will be very useful to the readers and to agricultural functionaries particularly extension functionaries in the country.

(J.P.Mishra)

13th November 2015   New Delhi

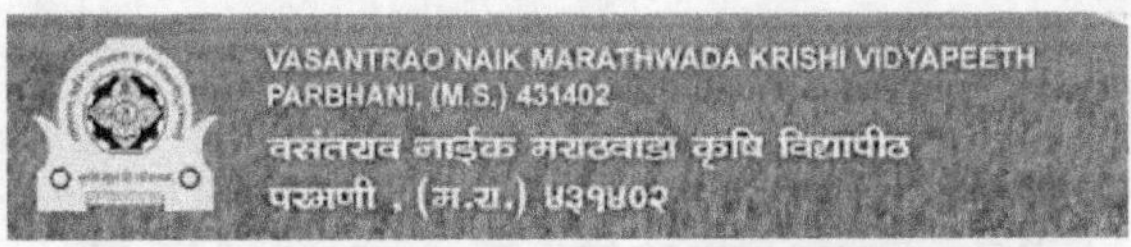

Dr. B. Venkateswarlu
Vice Chancellor

Foreword

Climate change is emerging as an important problem impacting all sectors of economy. Agriculture is one of the most vulnerable sectors in view of the high dependence of Indian Agriculture on monsoon and large number of resource poor farmers cultivating small holdings. Increasing temperatures and erratic rainfall are two major manifestation of changing climate. Several global climate models predict that the average temperatures in India are expected to rise by as high as of 4.5 ºC by end of this century. Warming causes melting of glaciers, increases demand for water by crops through more evaporation and transpiration and loss of soil organic carbon through rapid oxidation. The models also predict no major change in the total rainfall but considerable variation in the spatial and temporal distribution. The number of rainy days are likely to come down. In other words, the same rainfall is likely to fall in more intense storms with extended dry spells. Therefore rainwater management is going to be a major challenge for India to cope with changing climate. We need to manage dry spells and water logging due to heavy rainfall in a short period in the same season.

Climate change is not merely a technological issue. Tackling the impacts of climate change requires both technological and policy inputs besides social awareness. It is very important for all the stakeholders to understand its implications and contribute their best towards adaptation and mitigation. However there is very little awareness among general public and even officers of major technical and administrative departments at the state level on the drivers of climate change and the detailed understanding of programmes aimed at adaptation and mitigation. There are large number of programmes being implemented at Centre and state level on climate change adaptation. The aim of all these programmes is to enable Indian Agriculture cope with climate change, commonly known as "Climate Smart Agriculture".

The programme organized by the International Food Policy Research Institute (IFPRI) South Asia office, New Delhi and Centre for Good Governance (CGG), Hyderabad has fulfilled this important objective of capacity building on Climate Smart Agriculture. The publication brought out based on the presentations is quite comprehensive covering the fundamental aspects of climate change, vulnerability assessment, local and regional level adaptation and mitigation strategies, farmers knowledge/awareness and the related institutional, financial and policy issues including risk management through weather based crop insurance. I believe that this publications will be quite useful to all those interested in understanding Climate Smart Agriculture, plan and implement projects on adaptation and mitigation. I compliment the organizers of the workshop-IFPRI and CGG and the editor for the initiative in bringing out such a useful publication.

(B. Venkateswarlu)

Preface

Climate change poses many threats to agriculture, including the reduction of agricultural productivity and incomes particularly in India where over 80 percent farmers have small holdings with limited means of coping with adverse weather. Agriculture must undergo a significant transformation in order to meet the challenges of achieving food security and responding to climate change. The concept of Climate Smart Agriculture (CSA) was coined by Food and Agriculture Organization (FAO) and gaining increasing attention for sustainable productivity, resilience (adaptation), emissions reductions, national food security and development goals under one umbrella. The concept requires building competencies of agricultural extension functionaries for the promotion of climate adaptation concepts.

Accordingly a capacity building program on Climate Smart Agriculture was developed to build the capacity of senior and middle level extension functionaries working agriculture and allied departments in various states. This helps to improve their knowledge, attitudes to provide better services to farmers to become acquainted with opportunities and relevant technologies with respect to adaptation to climate change and Climate Smart Agriculture. The program is part of the efforts of International Food Policy Research Institute (IFPRI) that has recognised a need to provide guidance and support to address the effects of climate change on agriculture and the potential adaptation measures. The capacity building program is designed to provide an overview of the challenges that the agriculture sector faces due to climate change and to explore ways to address these challenges. The program applies different methods to transfer information, create substantial skills and enhance appropriation. Although it is designed for senior and middle level extension functionaries of agriculture departments in Karnataka, Andhra Pradesh and Telangana, it can be used as a training base for other audiences who are interested in climate resilient and environmentally sound agricultural production. The resource persons of the capacity building program provided useful materials to prepare this manual.

The publication responds to the needs of extension field functionalities and policy makers in agriculture and climate change scientists to understand and address better about sustainable food production dimensions in the era of climate change. It clarifies the linkages between agriculture and climate change mitigation and adaptation. It will help to build capacity among policy and decision-makers at global, regional and local level to design and implement agriculture related climate change policies, strategies and programmes, so that efforts to mitigate and adapt to climate change will help sustainable agricultural production.

This book provides climate change adaptation trainers and learners with a wealth of practice based adaptation knowledge and a process to asses, analyze and plan for focused climate change adaptation initiatives within their work profiles. The manual is sectioned into four training modules, setting a common understanding around climate change concepts, impact, global perspectives, then moving to more integrated knowledge regarding practices and technologies available towards climate smart agriculture and resilience building. The module, 'Financing Opportunities for Climate Smart Agriculture' deals with financial instruments and risk transfer mechanisms available in climate smart agriculture. Finally, a module helps agricultural administrators to plan and prepare climate smart agricultural projects under National Mission for Sustainable Agriculture.

Each module is comprised of several learning sessions, complete with handouts, participatory exercises, worksheets and learning aids prepared by experts in climate change and climate smart agriculture. Importantly, each session has a detailed step-by-step facilitation guide noting important concepts and knowledge to stress, and key information and advice on how to deliver the session. It is an advantage to the trainer or expert and participants that these modules be reviewed to provide further context and knowledge to enhance the learning experience in climate smart agriculture. The publication is being distributed at capacity building program   for participants from Karnataka, Andhra Pradesh and Telangana states. It will also help to become familiar with the establishment and monitoring of climate smart practices and to conduct training programmes related to climate smart agriculture for farmers.

This book is basically a guide for educators and training facilitators. It is often said that knowledge is power. But in changing climate scenario, access to information on climate smart agriculture is access to power. It can only be hoped that the publication of this book and sharing of the experiences and knowledge contained in this book will contribute to empowerment of participants. In order to clarify issues linked to climate change and agriculture, it has been necessary to write a book that seeks to cover many aspects on the subject. For the participants the training could be the first step to become a trainer for climate smart agriculture. The participants who would impart training at district level were encouraged to be interactive and participatory throughout the sessions and make this capacity building programme as their "own." This makes the participants to learn and pass their acquired knowledge, skills and attitudes to others.

-Editor

Acknowledgement

This volume of peer reviewed chapters which can be used as training manual arose from a capacity building programme on "Climate Smart Agriculture" conducted by Centre for Good Governance (CGG) at Hyderabad in association with International Food Policy Research Institute (IFPRI) during 29-31 July 2015. This programme brought together scholars and scientists working on climate smart agriculture who delivered training on topics of climate change, risks, mitigation and adaptation etc. I would like to express my sincere appreciation to all the resource persons and contributors to this book who brought a full array of approaches, frameworks, discourses and tools to help in understanding and begin to respond to the crucial relationship between climate change and agriculture. The contributions of the resource persons were reviewed, revised and updated while publishing the book.

I would like to thank International Food Policy Research Institute (IFPRI) for funding the capacity building programme on "Climate Smart Agriculture" that led to publishing this book. I wish to express my sincere gratitude to Dr.P.K.Joshi, Director-South Asia, IFPRI, New Delhi for his quality of communication and guidance. The excellent support provided by Ms.Vaishali Dassani, Program Manager of IFPRI is noted with thanks.

I am thankful to Dr.K.Padmaja, Asst Director of Agriculture on deputation to CGG, Mr.G.Kiran Kumar, Agriculture Consultant and Mr.A.Venkat Ramana, Project Associate at AMRG, CGG for patiently working on proposal stage to planning of the capacity building programme.

Special thanks are extended to my Director General Mr. K. Ramakrishna Rao, IAS for his encouragement and freedom provided to the oversight of the programme.

-Editor

1

Capacity Building in Extension: Key to Climate Smart Agriculture

Devi Prasad Juvvadi
Centre for Good Governance, Hyderabad.

1.1 Introduction

Currently India is staring at an agrarian crisis with daunting challenge of producing adequate food from an already shrinking natural resource to feed increasing population in the country. The pressing need therefore is to address this issue by increasing food grain production using methods that enhance productivity with efficient resource use by intensifying agricultural activities and to produce under conditions of variable climate.

It is well established in recent years that apart from natural resources and manmade physical capital, the human capital i.e., capacity of people as effective and productive change agents are becoming important for agricultural and rural development. This is supported by several studies that the education and skills of agricultural extension functionaries and farmers are significant factors in explaining the inter-farm differences in agricultural productivity, along with the more conventional factors such as availabilities of land and water resources, inputs, credit etc. (FAO, 1995).

After green revolution, many farming systems that were developed to be highly productive, have also gradually become susceptible to significant shocks such as drought, and more gradual changes such as climate change. These have increased the interest of scientists, policy makers and farmers in resilient farming systems and viewed as a necessary attribute to assist agricultural development through proper addressing future challenges and shocks. It may be noted that resilient farming systems are conceived as those which can cope with change and maintain productive capacity in the face of ongoing variability in factors such as commodity prices, climate, regulation and input availability (Chikaire et al., 2015).

Climate Smart agriculture (CSA) has emerged in recent years as a conceptual framework intended to align international efforts to increase agricultural productivity, mitigate greenhouse gas emissions, and reduce farmers' vulnerability to climate change. The FAO defines CSA as "agriculture that sustainably increases productivity, resilience (adaptation), reduces/removes greenhouse gases (mitigation) while enhancing the achievement of national food security and development goals" (FAO 2010). The concept of CSA has received impetus by a number of international multi lateral agencies such as the World Bank, the International Fund for Agricultural Development (IFAD), and the Consultative Group for International Agricultural Research (CGIAR). Additionally, the continuing development is furthering an agenda for policy and practice (Scherr, Shames, and Friedman 2012).

There is growing consensus among academicians, researchers and policy makers that climate change and its variability are emerging as the major challenges influencing the performance of Indian agriculture. Climate change increases the vulnerabilities to agriculture due to variability in temperature and rainfall and in the reduction of rainfall. Climate change variability also results in the deterioration of fresh water sources and increase flooding which could in turn adversely affect agricultural yields. The case of India is especially significant with several climate extreme weather events being seen such as frequent droughts, floods, cyclones, hailstorm, frost etc. It also becomes significant given that more than 70% of India's rain requirements are dependent on changes in the climate and the excessive dependence of more than 75 million people on agriculture and allied livelihoods as this creates increased pressures on the available natural resources and coping mechanisms. Evidence already points to climate change based vulnerabilities in India leading to reduced economic growth and an accompanying inflation of essential food commodities which severely threatens the food security of the country.

All this calls for a Climate Smart Agriculture (CSA) leading to sustainable food security through integrating innovations, technologies, efficient resource use, sound public policies, establishment of new institutions, and development of infrastructure. The development of new technologies and their diffusion and management of capabilities for more intensive agriculture and supporting services become imperative in climate smart agriculture. This can be achieved through agricultural extension in agriculture and allied sectors.

1.2 Preparedness of Extension for CSA

Farmers need timely and location specific information on climatic risks but currently there are a number of gaps and challenges in providing climate information to the farmers. First amongst them is non-preparedness of extension functionaries in terms of climate change. Agriculture and allied departments and extension organisations in many states are completely unaware of climate change impacts on agriculture. Preparedness like documenting climate change scenarios at grassroots level, extent of adaptation (individual/farmer group level), mapping vulnerable regions, sustainable indicators, access to real time data, effective synthesis & interpreting, better decision making for a climate change scenario etc., are missing at present. The agriculture and allied departments and extension organisations consider climate change as seasonal aberrations. Integrating the multi-disciplinary and multi-sectoral information into a meaningful extension material is beyond the expectations of traditional extension systems. The agricultural extension systems also need to be geared up to take upon the challenge.

1.3 Capacity Building

Hilderbrand (2008) described, capacity is the mean or the ability, to fulfil a task or meet an objective effectively. It refers to the skills of staff and strength of specific organization and thus, training staff and strengthening organization is called capacity building.

Capacity building means a new build-up of capabilities (Kogut, 1992) and refers to activities that improve an organization's ability to achieve its mission or a person's ability to define and realize his/her goals or to do his/her job more effectively. Capacity building is as important as capital investment and infrastructure (Mati, 2008) since it also increases the abilities and resources of persons, communities and organizations to manage change (Coutts et. al., 2005).

UNESCO (2006), reports that capacity building focuses on increasing an individual and organization's abilities to perform core functions, solve problems, and objectively deal with developmental needs. Furthermore, capacity building is understood to be resulting in improving or upgrading the ability of the person, team and institutions to implement their functions and achieve goals over time and important for all levels, from individuals to national organizations (Horton, 2002). In addition to improving individuals and teams, capacity building also refers to enhancing the organizational capacities of communities, and is crucial for formation of non-profit organizations (Paul and Thomas, 2002).

1.4   Importance of Capacity Building in Climate Smart Agriculture

Capacity building is essential because it improves knowledge and skills to keep pace with the rapidly changing technological developments and also helps improve the feedback mechanisms from the field to the extension staff and researchers (Dwarakinath, 2006). Capacity building is needed as it helps in not just farmer training and strengthening the innovation process but also in building linkages between farmers and the various stakeholders involved in helping farmers (Sulaiman, 2006). It is generally reported that public extension services often have some weaknesses such as lack of timely information and input supply, less accountability of public extension personnel, the blanket nature of recommendations, and the absence of extension personnel during office hours, which make them less committed to the service.

The extension capacity building is often overlooked in the rush to push the results of research and development products to farmers. Building the capacity of agricultural extension staff is central to success of climate smart agriculture initiatives. Farmers in many developing countries including in India are resource constraint, poor and still have weaknesses in their development. Agriculture and allied departments are already pressurized with their regular set of activities and do not focus much on the emerging challenge of climate change. Therefore, improving the capacity building of agriculture extension is necessary for climate smart agriculture, poverty alleviation and environment protection. Implementing climate smart agriculture projects alone cannot lead to the desired level of development in agriculture without building the capacity of farmers, groups and organizations because they must have the ability and responsibility to resolve their problems and develop their communities.

The small and marginal farmers must be encouraged for participation and giving opportunities. Along with men, women play an important role in contributing to all activities in agricultural operations such as livestock production, fisheries, cropping, forestry, irrigation and horticulture. The extension staff working closely with farmers should explain how climate change affects their lives and incomes not only from agriculture but from livestock etc. The extension workers should be in a position to introduce appropriate technologies or information relevant to specific farmer situation like livestock, horticulture etc. There are four main tools for the development of capacities; information dissemination, training, facilitation and mentoring, networking and feedback to promote learning from experience with each having advantages and disadvantages (Horton, 2002).

Chikaire et al (2015) described that training is often used as the main capacity building method for agricultural extension in developing countries including on-the-job training and workshops etc that can be applied to climate smart agriculture as well. He also reported that of central importance to most capacity building is "learning-by-doing" . The learning-by-doing approach has been an important part of education to develop capacity and insights in a wide range of settings. Learning-by-doing is one of the most commonly quoted processes through which partners☐ capacities are understood to develop, and it is a good way for people to learn. Individuals, groups, and some organizations can learn-by-doing (Gillespie, 2004). Learning-by-doing or experiential learning is at the heart of capacity development.

Experiential learning is an excellent way to develop good insights into the ways groups work (Banjarmin et al., 1997) and can be a very powerful method and appropriate in extension for climate smart agriculture particularly who work with groups. Demonstration plots, cross visits, study tours and Farmer Field School are useful methods to transfer information and technology to staff and farmers, particularly in remote areas. The advantages of farmer field schools (FFS) are that both farmers and staff are able to gain knowledge, skills, good relationships, facilitator skills, communication skills and experiences.

Mentoring is an important method for capacity building in extension and should be wisely used for climate smart agriculture. Mentors are senior research and extension staff who are experienced persons. Mentors are people who have more experience in indigenous technical knowledge (ITK) and extension methodology. Mentoring involves passing on skills, attitudes and knowledge from experienced staff to newer extension workers. Millar & Connell (2005) stated that building the technical and extension skills of staff using experienced people as mentors is a key element of scaling out impacts. They can provide the support trainees need in order to become responsible as they acquire new skills and adapt to change. Mentors should be highly skilled in communicating, listening, analyzing, providing feedback and negotiating with less experienced persons.

1.5   Need for Capacity Building in CSA

Capacity-building in agriculture and allied sectors primarily addresses the establishment or strengthening of formal (government) and informal (NGOs, farmer groups, etc.) institutions, the private sector and individuals. The aim is to enable them to be better able to face their responsibilities in policy and   decision-making   and   in   implementing   agricultural development programmes more efficiently. This also implies decentralization down to local level, and providing incentives for local community initiatives and people's participation. It is necessary that voluntary organizations and those representing the interests of the various farmer interest groups should be involved to have a capacity to train farmers, cooperatives and other rural organizations to assist in consolidating grassroots organizations. In-service training of extension functionaries of agriculture and allied departments in participatory techniques is an essential complement to the involvement of local groups.

Climate smart agriculture places the capacity building challenge for the extension profession in to the domains of multi and inter-disciplinary work (Chikaire et al., 2015). The question that emerges from such challenges is: how can the capacity of extension match the challenge of supporting learning and resilience in climate smart agriculture across these domains?

The agricultural and allied departments generally lack the full range of in-house expertise to respond to the changes required for sustainable agriculture and climate smart agriculture. Ongoing capacity building and training programmes to extension functionaries should include modules to enhance knowledge and skills in climate smart technologies. There are several sectors and actors are involved in climate smart agriculture and in planning for climate smart agriculture, the capacities of actors involved in different sectors need to be enhanced, particularly at local level, where actions are needed. Agriculture officers and extension field functionaries are a vital link in the translation of knowledge obtained from research into on field adoption. The extension field functionaries are fundamental in agriculture development at field level. These people need to fully understand problems, needs and possible ways to promote climate smart agriculture, so that they can communicate to farmers. Hence, agriculture department officials and extension field functionaries, more than, ever will be the catalysts of a climate smart agriculture. Change at local levels can bring a marked improvement in the amelioration of undesirable effects of climate change. It is vital that capacities of these staff are built on a topic that has a global effect and each small effort is crucial. Middle and senior level Agriculture

Extension officials with their experience and expertise are ideally suited for this Capacity Development Programme as training of trainers.

Ozor and Nnaji (2011) proposed four key roles for extension in agricultural adaptation to climate change, namely;

- Training and re-training of extension staff to acquire new capacity in climate change management
- Setting up of emergency management units in extension agencies
- Dissemination of innovations on best adaptation practice
- Improving feedback to government and interested agencies on climate change issues

Essentially, the knowledge base of extension functionaries on climate change should be expanded, so that they in turn will be able to train farmers. Despite the existence of several adaptation strategies that have been proven, little to no attempts have been made to develop appropriate training curriculum that incorporates the various adaptation strategies. Hence, there is a need to enhance the capacity of extension functionaries on climate change issues such as causes of climate change, its effects and adaptation strategies (Singh and Grover 2013).

1.6   Areas of Competence for Extension Functionary in CSA

Chikaire et al., (2015) reported that extension functionaries must demonstrate sufficient competence that includes:

- **Communication:** The extension functionaries must be able to convey agricultural information to all categories of farmers rich and poor, learned and illiterate, as well as possess the disposition to mildly persuade them to adopt innovations.
- **Farming:** The change agent must be able to demonstrate new technologies to the farmers even if involves physical work and practice.

- **Science:** The ability to read and understand professional literature as well as the ability to carry out field experiments is needful assets for the extension functionary.

- **Economics:** The change agent must be able to analyze and recommend cost-benefit strategies based on knowledge of prevailing market situations, agricultural policies, availability of credit, cost-benefit ratio, interests, etc.

- **Social:** The extension functionary must be familiar with the customs, values and ways of thinking of the farmers as to work in tandem with the realities of the people and thus avoid socio-cultural conflict.

The extension functionaries in climate smart agriculture must also possess basic competencies such as congruency, empathy and appreciation along with necessary technical and methodological competencies. These include being credible and knowledgeable in the subject matter along with being able to use specific communication techniques, appropriate media and communication aids. Managerial and organizational competencies such as being able to work within the framework of facilitation and guidance are also imperative.

Information Communication Technologies (ICTs) is improving and changing how extension work is carried out in agriculture and allied sectors. The use of electronic media for agricultural extension for climate smart agriculture can assist extension agents reach the farmers across different location with location specific information. Entrepreneur extension agents can record and produce videos in local dialects with local farmers featuring in the videos thereby making it easier for farmers to understand. Capacity building is needed in the area of agricultural information so that farmers are able to identify and replicate best practices without any external interventions. Specifically, it is needed in computer multimedia production, video production, E-journalism, photo journalism, etc. Gender mainstreaming in extension, entertainment education for extension work, climate change, among other evolving issues in extension also offer need dimensions to capacity building in extension (Chikaire et al., 2015) for climate smart agriculture.

Farmers on their own can identify the practices as authentic and replicable in their context and package same in Double Video/Compact Disc (DVDs/CDs) or uploaded online for farmers and other users of agricultural information achieve these and other uses of ICTs. Capacity is therefore needed in computer multimedia production, video production, E-journalism, photo journalism, etc, in order to bring the earlier barriers of research information dissemination through electronic media collapsing.

1.7   Objectives of the Capacity Building for CSA

Capacity building in climate smart agriculture should incorporate practices that increase productivity, efficiency, resilience, adaptive capacity, and mitigation potential of production systems. The capacity building will have to address how to integrate climate change, water, and natural resources management into existing and proposed agricultural and food security programs to increase productivity, sustainability of investments, and climate resilience.

Improving knowledge of key factors in climate smart agriculture is critical for initiating field interventions and the success of agriculture because it is a broad and multi faceted issue (Bogdanski, 2012). There are many barriers for the development and implementation of climate smart agriculture initiatives at field. Some examples include the slow adoption of technological innovations, regulatory and policy issues, lack of demand or the unequal distribution of benefits and costs across supply chains (Blok et al., 2015).

The challenges associated with climate smart agriculture will require, among other aspects, integrated multi-disciplinary approaches, which build upon social, natural and technological sciences. The modules in capacity building programmes should address the following issues:

- What are the best methods for promoting climate smart agriculture and what strategies exist for expanding or up scaling successful examples of climate smart agriculture?

- How can socio-economic, political, ecological and technological approaches be combined and integrated to help overcome key barriers to climate smart agriculture?

- What concepts, approaches or actions are available for advancement of climate smart agriculture?

- How can costs and benefits of climate smart agriculture be spread in agricultural production?

- What successful examples exist of climate smart agriculture innovation, adoption and diffusion?

- What role should policy play in furthering climate smart agriculture?

- What are the key topics and priorities for a successful near and long-term future research agenda for climate smart agriculture?

Conclusions and Recommendations

Capacity building initiatives in Climate Smart Agriculture have began in the past few years, mostly on separate basis. Time has come to draw lessons from initial initiatives and to build on them to strengthen our knowledge basis about the learning objectives to be defined, about the methods of learning that can address these objectives and about the curricula and training tools that can be associated to them. Addressing adaptation or mitigation as a training objective cannot be fulfilled in isolated disciplines but has to be integrated with ongoing extension activities.

Consequently, there is an urgent need to pass right message about climate change by the extension agents to farmers. This will help farmers;

- To understand why climate change issues are so important in their daily activities

- Put together adjustment mechanisms to cope with this climate change phenomenon

- Reduce agricultural production losses related to climate change and in return become less vulnerable to climate change

- Put together clear adaptation strategies aimed at enhancing adaptive capacity, generate income and improve livelihoods.

As the climate change is emerging as new challenge for agriculture, the needs of farmers are changing and warrants for extension needs to periodically upgrade in knowledge, skills and attitudes in order to keep pace with the emerging challenges and dynamics of extension work in climate smart agriculture. Capacity building is essential in ensuring that the initial job training is provided in technologies available for coping climate change in agriculture and allied sectors and as well as ensuring coping to the job changes and the varied needs of the farming community. The capacity building programmes in CSA should explore location specific natural resources and water management issues critical to agriculture, identify challenges of climate variability and change, and examine practices that build improved resilient and productive agricultural systems. The, extension functionaries in agriculture and allied departments are the agents of change in fostering development in agriculture, animal husbandry, horticulture, fisheries etc. Therefore, the sooner extension and other service providers become familiar with climate change, the earlier the integration of climate change into agricultural developmental goals.

References

Benjamin J., Bessant J., and Watts R. (1997). Making Groups Work: Rethinking Practices, Allen and Unwin, St. Leonards. 1997.

Blok, V., Long, T.B., Coninx, I. (2015). Barriers to the adoption and diffusion of CSA technological innovations in Europe, Climate Smart Agriculture: Global Science Conference, March 16th - 18th, Montpellier, France.

Bogdanski, A. (2012). Integrated food-energy systems for climate-smart agriculture. Agriculture & Food Security 1, 9.

Chikaire, J.U., Ani, A.O., Atoma, C.N. and Tijjani, A.R. (2015) Capacity Building: Key to Agricultural Extension Survival . *Scholars Journal of Agriculture and Veterinary Sciences.2015; 2(1A):13-21.*

Coutts J., Roberts K., Frost F. and Coutts, A. (2005). The Role of Extension in Building Capacity-What Works, and Why. 2005. Available at www.fao.org./sd/exdirect/Exan0015.htm

Dwarakinath, R. (2006). Changing Tasks of Extension Education in Indian Agriculture. In A.W. Vanden Ban, and R.K. Samanta, (Eds.), Changing

Roles of Agricultural Extension in Asian Nations (pp. 56-79). Delhi: B.R. Publishing, India.

FAO (1995). World Agriculture Towards 2010: An FAO Study. Food and Agriculture Organization, Rome. 1995.

FAO (2010). Climate-Smart Agriculture - Policies, Practices and Financing for Food Security, Adaptation and Mitigation.

Gillespie S (2004). Scaling up community-driven development: A synthesis of experience. Washington, DC: World Bank Social Development Paper, 2004. 69.

Hilderbrand, M.E., (2008). Capacity for Poverty Reduction: Reflection on Evaluations of System Efforts: IN Capacity Building for Poverty Eradication Analysis of, and Lessons from Evaluations of UN System Support to Countries; Efforts. United Nations, New York.

Horton, D. (2002); Planning, Implementing and Evaluating Capacity Development. ISNAR Briefing Paper 50. July 2002.Available at http://citeseerx.ist.psu.edu/viewdoc/ download?doi=10.1.1.518.8285 &rep=rep1&type=pdf

Kogut. B. and Zander, U. (1992). Knowledge of the firm, combinative capabilities, and the replication of technology. Organization science, 1992; 3(3):383-397.

Mati, B.M. (2008). Capacity Development for Smallholder Irrigation in Kenya. International Crops Research Institute for Semi-Arid Tropics, Kenya, Nairobi. 2008.

Millar.J., Photakoun. V., and Connell. J. (2005). Scaling Out Impacts: A Study of Three Methods for Introducing Forage Technologies to Villages in Lao PDR. Australian Center for International Agricultural research, Australia. 2005.

Ozor, N. and Nnaji, C. (2011). The Role of Extension in Agricultural Adaptation to Climate Change in Enugu State, Nigeria. Journal of Agricultural Extension and Rural Development, vol. 3 (3), pp. 42-50.

Paul, C. and Thomas,. (2002). W; The Participatory Change Process: A Capacity Building Model from a US NGO. Development in Practice, 2002; 10 (2): 240 – 244.

Scherr, S. J., Shames, S., and Friedman, R. (2012). From Climate-Smart Agriculture to Climate-Smart Landscapes. Agriculture & Food Security, 1, 12.

Singh, Iqbal and Jagdish Grover. (2013): Role of extension agencies in climate change relatedadaptation strategies. *International Journal of Farm Sciences* 3(1) : 144-155, 2013.

Stephen. P, Brien. N. and Triraganon, R. (2006). CapacityBuilding for CBNRM in Asia: A Regional Review. 2006. Available at www.elcoftc.org/site/fundmm/dois/CABS/idrc-review.pdf

Sulaiman, V.R. and Hall, A. (2006). Extension Policy Analysis in Asian Nations. In Vanden Ban, A.W. and Samanta, R.K. (eds) Changing Roles of Agricultural Extension in Asian Nations. B.R. Publishing Corporation, India.

2

Basic Concepts of Climate Change with Global and Local Perspectives

Ch. Srinivasa Rao and Arun K. Shankar
*ICAR Central Research Institute for Dryland Agriculture (CRIDA),
Hyderabad.*

2.1 Introduction

Climate change is a long-term shift in the statistics of the weather (including its averages). For example, it can be seen as a change in climate normal (expected average values for temperature and precipitation) for a given place and time of year, from one decade to the next. It is has been found that the global climate is currently changing. The last decade of the 20th Century and the beginning of the 21st have been the warmest period in the entire global instrumental temperature record, starting in the mid-19th century. Enormous progress has been made in increasing our understanding of climate change and its causes, and a clearer picture of current and future impacts is emerging. Research is also shedding light on actions that might be taken to limit the magnitude of climate change and adapt to its impact. There is consensus among the majority of climate scientists agree that human activities, especially the burning of fossil fuels (coal, oil, and gas), are responsible for most of the climate change currently being observed.

2.2 Global Climate Change: Concepts and Effects

2.2.1 What is Climate Change?

Before we go into what exactly is climate change it is important to understand what the difference between weather and climate is. Many people use the words weather and climate as if they mean the same thing, but they are not the same. It is important to understand the difference in order to appreciate the idea of climate change. Weather is what is going on in the atmosphere at a particular place and time. Weather is measured in terms of wind speed, temperature, humidity, atmospheric pressure, cloudiness, and precipitation. In most places, weather changes from hour-to-hour, day-to-day, and season-to-season. Climate on the other hand is the weather conditions prevailing in an area in general or over a long period. The word climate refers to the average pattern of weather in a region.

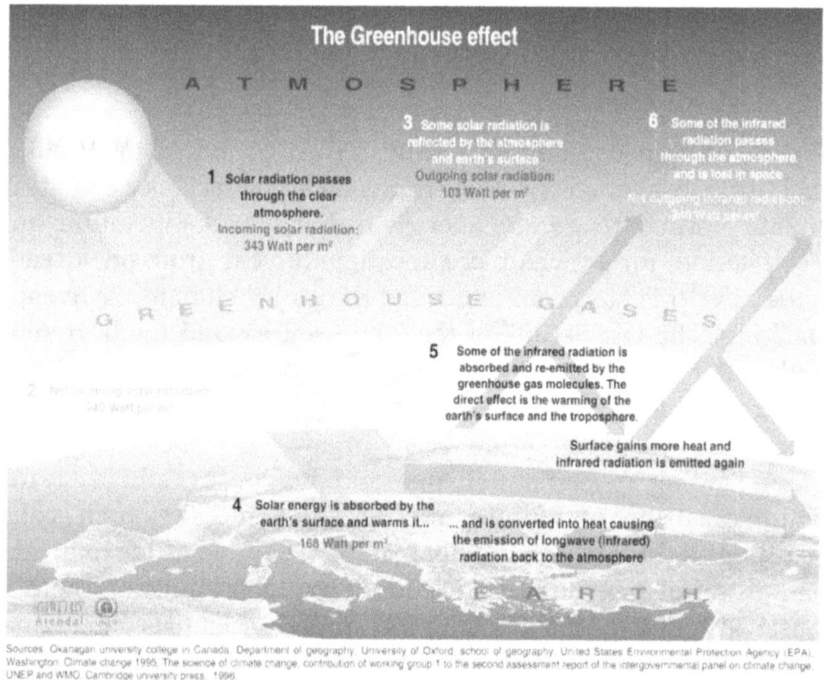

Sources: Okanagan university college in Canada. Department of geography. University of Oxford. school of geography. United States Environmental Protection Agency (EPA). Washington: Climate change 1995. The science of climate change, contribution of working group 1 to the second assessment report of the intergovernmental panel on climate change. UNEP and WMO. Cambridge university press, 1996.

Climate change denotes long term changes in climate including mean temperature and precipitation. Shifting weather patterns results in changing climate, there is sort of change threatens food production through increased unpredictability of precipitation, rising sea levels contaminate coastal freshwater reserves and increase the risk of catastrophic flooding. A warming atmosphere aids the pole-ward spread of pests and diseases once limited to the tropics. Green house effect is one of chief causes of climate change on Earth. The greenhouse effect is a process by which thermal radiation from earth's surface is absorbed by atmospheric greenhouse gases (GHG) and is re-radiated in all directions, the main greenhouse gases in the Earth's atmosphere are water vapour, carbon dioxide, methane, nitrous oxide and ozone. Since part of this re-radiation is back towards the surface, it results in an elevation of the average surface temperature above normal. The greenhouse effect plays an important role in regulating the climate of the earth. In the absence of this greenhouse effect earth's surface temperature would be far below freezing. However, an increase of these GHGs could result in increased trapping of heat and rising global temperatures.

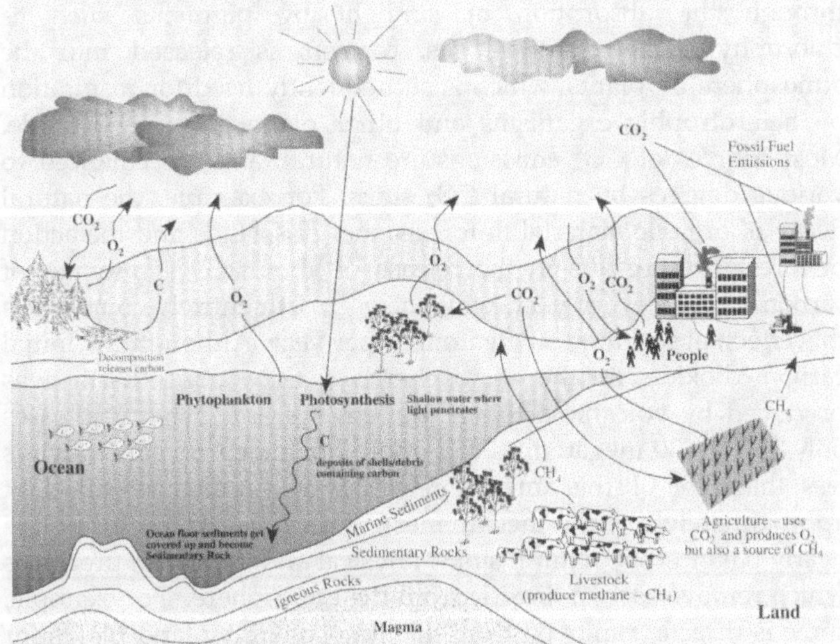

2.2.2 Carbon Dioxide the Primary Culprit

Carbon dioxide (CO_2) is one of the important GHGs which contribute to this change in climate. The concentration of carbon dioxide (CO_2) in Earth's atmosphere has reached 395 ppm (parts per million) by volume as of June 2012. The current level is the highest in the past 650,000 years. The Fourth Assessment Report of the Intergovernmental Panel on Climate Change concludes, "that most of the observed increase in the globally averaged temperature since the mid-20th century is very likely due to the observed increase in anthropogenic greenhouse gas concentrations." Natural sources of atmospheric carbon dioxide include volcanic outgassing, the combustion of organic matter, wildfires and the respiration processes of living aerobic organisms; man-made sources of carbon dioxide include the burning of fossil fuels for heating, power generation and transport, as well as some industrial processes. It is also produced by various microorganisms from fermentation and cellular respiration. Plants convert carbon dioxide to carbohydrates during a process called photosynthesis. They gain the energy needed for this reaction through the absorption of sunlight by pigments such as chlorophyll. The resulting gas, oxygen, is released into the atmosphere by plants, which is subsequently used for respiration by heterotrophic organisms and other plants, forming a cycle. Most sources of CO_2 emissions are natural, and are balanced to various degrees by natural CO_2 sinks. For example, the natural decay of organic material in forests and grasslands and the action of forest fires results in the release of about 439 gigatonnes of carbon dioxide every year, while new growth entirely counteracts this effect, absorbing 450 gigatonnes per year. Although the initial carbon dioxide in the atmosphere of the young Earth was produced by volcanic activity, modern volcanic activity releases only 130 to 230 megatonnes of carbon dioxide each year, which is less than 1% of the amount released by human activities (at approximately 29,000 megatonnes). These natural sources are nearly balanced by natural sinks, physical and biological processes which remove carbon dioxide from the atmosphere. For example, some is directly removed from the atmosphere by land plants for photosynthesis and it is soluble in water forming carbonic acid. There is a large natural flux of CO_2 into and out of the biosphere and oceans. In the pre-industrial era these fluxes were largely in balance. Currently about 57% of human-emitted CO_2 is removed by the biosphere and oceans. The ratio of the increase in atmospheric CO_2 to emitted CO_2 is known as the airborne fraction; this varies for short-term averages but is typically about 45% over longer (5 year) periods. Estimated carbon in global terrestrial vegetation increased from approximately 740 billion tons in 1910 to 780 billion tons in 1990. While CO_2 absorption and release is always happening as a result of natural processes, the recent drastic rise in CO_2 levels in the atmosphere is known to be entirely due to human activity such as burning fossil fuels like coal and petroleum which is the leading cause of increased anthropogenic CO_2; deforestation is the second major cause, as a result, carbon dioxide has gradually accumulated in the atmosphere, and as of 2009, its concentration is 39% above pre-industrial levels. Carbon dioxide has deleterious long-term effects on climate change that are largely irreversible; hence it is important to check the increase in CO_2 concentration in order to combat climate change.

2.3 Natural Variability and Climate Change

The climate varies naturally from one year to the next. Climate can also vary on much longer timescales, from decades to centuries. Climate change refers to the shift in the mean state of a particular climate parameter, such as temperature or precipitation. Natural climate variability refers to the variation in climate parameters caused by other than human forces. There are two types of natural variability the first type are those that are external and those that are internal to the climate system. Variations in the Sun, volcanic eruptions, and changes in the orbit of the Earth around the Sun exert an external control on climate variability. These processes are the driving force behind changes that occur over long time periods, such as oscillations between ice ages and interglacial periods. On the other hand, natural variability is also influenced by processes internal to the climate system that arise, in part, from interactions between the atmosphere and ocean, such as those occurring in the tropical Pacific Ocean during an El Niño event. These changes occur over shorter time periods, from months to decades. In any given year, natural variability may cause the climate to be different than its long-term average. The Intergovernmental Panel on Climate Change (IPCC) uses the phrase "climate change" to refer to any climate change that has occurred or will occur, whether from natural variability or human activity. In popular use, however, the phrase is often synonymous with the phrase "global warming," which is used to describe the world's ongoing temperature rise as it relates to society's emissions of greenhouse gases.

2.4   Interconnected Factors in Climate Change

The climate system is interconnected and contains relationships in which an event in one geographic area causes changes in another area, which may be a log distance away. The best example for this is the El Nino effect. These long distance interconnections are a natural consequence of the chaotic atmospheric system but have a recurring pattern that can be observed over periods of weeks to years. Although many interconnections have been recognized, combinations of only a small number of patterns can account for much of the interannual variability in the climate. These natural cycles create variability in the weather and climate that at times may accentuate the warming caused by the enhanced greenhouse effect or may suppress, and even override, human-caused warming. El Niño-Southern Oscillation (ENSO) is the most important source of interannual connections across the globe and causes large changes in climate. It is characterized by a temperature change in tropical Pacific sea surface temperatures that alters the strength and direction of atmospheric trade winds. In addition to this radiative effects are also envisaged. Radiative effects measures the influence that climate-altering factors have on the energy balance of the Earth. Examples of factors that can alter the Earth's energy balance include atmospheric concentrations of greenhouse gases, aerosols from volcanoes and air pollution, and the amount of solar radiation delivered to the Earth by the Sun.

2.5   Observed Climate Change Impacts

Rising temperatures due to increasing greenhouse gas concentrations have produced distinct patterns of warming on Earth's surface, with stronger warming over most land areas and in the Arctic. There have also been significant seasonal differences in observed warming. Global warming is also having a significant impact on snow and ice, especially in response to the strong warming across the Arctic. It has been observed that snow and ice are melting and frozen ground is thawing, hydrological and biological systems are changing and in some cases being disrupted, migrations are starting earlier, and species' geographic ranges are shifting towards the poles. For terrestrial biological systems, changes documented in the database include shifts in spring events (e.g., earlier leaf unfolding, blooming date, migration, and timing of reproduction), species distributions, and community structure. Because CO_2 reacts in seawater to form carbonic acid, the acidification of the world's oceans is another certain outcome of elevated CO_2 concentrations in the atmosphere. Extreme weather includes unusual, severe or unseasonal weather; weather at the extremes of the historical distribution—the range that has been seen in the past. Often, extreme events are based on a location's recorded weather history and defined as lying in the most unusual ten percent. In recent years some extreme weather events have been attributed to human-induced global warming, with studies indicating an increasing threat from extreme weather in the future. Although it is difficult to attribute the cause of any one event to global warming, there is solid theoretical justification for expecting the frequency and/or magnitude of some extreme events to increase in the future. It is also likely that while some events will become more common and powerful, others will become less frequent and intense, like in the temperature example mentioned above.

2.6   Local Climate Change

The effects of global warming on the Indian subcontinent vary from the submergence of low-lying islands and coastal lands to the melting of glaciers in the Indian Himalayas, threatening the volumetric flow rate of many of the most important rivers of India and South Asia. In India, such effects are projected to impact millions of lives. As a result of ongoing climate change, the climate of India has become increasingly volatile over the past several decades; this trend is expected to continue. Climate change is impacting the natural ecosystems and is expected to have substantial adverse effects in India, mainly on agriculture on which 58 per cent of the population still depends for livelihood, water storage in the Himalayan glaciers which are the source of major rivers and groundwater recharge, sea-level rise, and threats to a long coastline and habitations. Climate change will also cause increased frequency of extreme events such as floods, and droughts.

2.6.1 Extreme Heat

India is already experiencing a warming climate, unusual and unprecedented spells of hot weather are expected to occur far more frequently and cover much larger areas. Under 4°C warming, the west coast and southern India are projected to shift to new, high-temperature climatic regimes with significant impacts on agriculture.

2.6.2 Changing Rainfall Patterns

A decline in monsoon rainfall since the 1950s has already been observed. The frequency of heavy rainfall events has also increased. A 2°C rise in the world's average temperatures will make India's summer monsoon highly unpredictable. At 4°C warming, an extremely wet monsoon that currently has a chance of occurring only once in 100 years is projected to occur every 10 years by the end of the century. An abrupt change in the monsoon could precipitate a major crisis, triggering more frequent droughts as well as greater flooding in large parts of India. India's northwest coast to the south eastern coastal region could see higher than average rainfall. Dry years are expected to be drier and wet years wetter.

2.6.3 Droughts

Evidence indicates that parts of South Asia have become drier since the 1970s with an increase in the number of droughts. Droughts have major consequences. In 1987 and 2002-2003, droughts affected more than half of India's crop area and led to a huge fall in crop production. Droughts are expected to be more frequent in some areas, especially in north-western India, Jharkhand, Orissa and Chhattisgarh. Crop yields are expected to fall significantly because of extreme heat by the 2040s.

2.6.4 Groundwater

At 2.5°C warming, melting glaciers and the loss of snow cover over the Himalayas are expected to threaten the stability and reliability of northern India's primarily glacier-fed rivers, particularly the Indus and the Brahmaputra. The Ganges will be less dependent on melt water due to high annual rainfall downstream during the monsoon season. The Indus and Brahmaputra are expected to see increased flows in spring when the snows melt, with flows reducing subsequently in late spring and summer. Alterations in the flows of the Indus, Ganges, and Brahmaputra rivers could significantly impact irrigation, affecting the amount of food that can be produced in their basins as well as the livelihoods of millions of people (209 million in the Indus basin, 478 million in the Ganges basin, and 62 million in the Brahmaputra basin in the year 2005).

2.7   Climate Change and Agriculture

Climate change and agriculture are interconnected processes, both of which take place on a global scale. Agriculture is particularly vulnerable to climate change. Higher temperatures will tend to reduce yields of many crops; it may encourage weed and pest proliferation. Changes in precipitation patterns increase the probability of short-run crop failures and long-run production falloffs. Although there will be increases in some crops in some regions of the world, the overall impacts of climate change on agriculture are expected to be negative, threatening global food security. In developing countries, climate change will cause yield declines for the most important crops. South Asia will be particularly hard hit. Climate change will have varying effects on irrigated yields across regions, but irrigated yields for all crops in South Asia will experience large declines. Climate change will result in additional price increases for the most important agricultural crops – rice, wheat, maize, and soybeans. Higher feed prices will result in higher meat prices; as a result, climate change will reduce the growth in meat. In India the impact of climate change on agriculture is expected to be more severe than realized earlier, in particularly in wheat if there is one degree rise in temperature in areas like Uttar Pradesh, Punjab and Haryana could amount to a loss of about several million tonnes annually, in addition production in   crops such as rice, millet and maize could reduce up to 10 percent. Decrease in potential yields is likely to be caused by shortening of the growing period, decrease in water availability and poor vernalization. Biodiversity is also adversely affect by climate change which in turn affects agricultural production, this is particularly important to the marginal and small farmers of India. Poor people, especially those living in areas of low agricultural productivity, depend especially heavily on the genetic diversity of the environment. The effective use of biodiversity at all levels - genes, species and ecosystems - is therefore a precondition for sustainable development.

Projected impact of climate change on agricultural yields

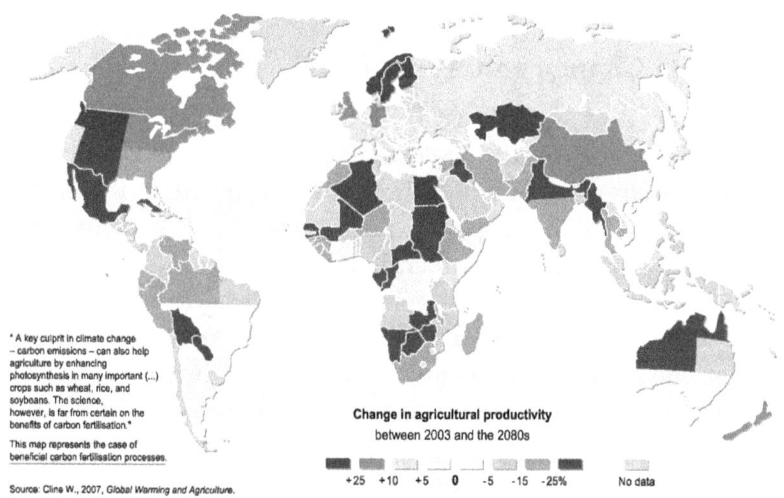

It is important to assess the effects of global climate changes on agriculture this will help to properly anticipate and adapt farming to maximize agricultural production in a sustainable way. The consequence of climate on agriculture is related to variabilities in local climates rather than in global climate patterns. The Earth's average surface temperature has increased by 0.83°C since 1880. Therefore, it is important for any assessment to be done individually considering each local area. It is important to note that not all effects of climate change are negative, favourable effects on yield can be seen by realization of the potentially beneficial effects of carbon dioxide on crop growth and increase of efficiency in water use.

The main effects of climate change in agriculture are the following

- Productivity, in terms of quantity and quality of crops
- Agricultural practices, through changes of water use and agricultural inputs such as herbicides, insecticides and fertilizers
- Environmental effects, in particular in relation of frequency and intensity of soil drainage (leading to nitrogen leaching), soil erosion, reduction of crop diversity.

Conclusions

Even though climate change in India is now a reality, a more certain assessment of the impacts and vulnerabilities of agriculture sectors including livestock and fisheries is needed. A comprehensive understanding of adaptation options across the full range of warming scenarios and regions would go a long way in preparing the nation for climate change. A multi pronged strategy of using indigenous coping mechanisms, wider adoption of the existing technologies and/or concerted R&D efforts for evolving new technologies are needed for adaptation and mitigation. Policy incentives will play crucial role in adoption of climate ready technologies in all sectors. The state agricultural universities and regional research centres will have to play major role in adaptation research which is more region and location specific while national level efforts are required to come up with cost effective mitigation options, new policy initiatives and global co-operation.

References

IPCC, 2013. Climate Change 2013: The Physical Science Basis. India: Climate Change Impacts World Bank Report June 2013. Contribution of Working Group I to the Fifth Assessment Report of the Intergovernmental Panel on Climate Change [Stocker, T.F., D. Qin, G.-K. Plattner, M. Tignor, S.K. Allen, J. Boschung, A. Nauels, Y. Xia, V.

Bex and P.M. Midgley (eds.)]. Cambridge University Press, Cambridge, United Kingdom and New York, NY, USA, 1535 pp, doi:10.1017/CBO9781107415324.

Hansen, G., and Cramer, W. (2015). Global Distribution of Observed Climate Change Impacts. Nature Climate Change, 5(3), 182-185.

Rupa Kumar, K., Sahai, A. K., Krishna Kumar, K., Patwardhan, S. K., Mishra, P. K., Revadekar, J. V., & Pant, G. B. (2006). High-resolution Climate Change Scenarios for India for the 21st century. Current science, 90(3), 334-345p.

Srinivasarao, Ch., Venkateswarlu, B., Rattan Lal, Singh, A. K. and Kundu, S., (2013). Sustainable Management of Soils of Dryland Ecosystems of India for Enhancing Agronomic Productivity and Sequestering Carbon, In: *Advances in Agronomy* (Ed. Donald L. Sparks), Academic Press, Burlington, 121, 253-329p.

Srinivasarao, Ch., Venkateswarlu, B., Sikka, A. K., Prasad, Y. G., Chary, G. R., Rao, K. V., Gopinath, K. A., Osman, M., Ramana, D. B. V., Maheswari, M. and Rao, V. U. M., (2015). District Agriculture Contingency Plans to Address Weather Aberrations and for Sustainable Food Security in India, ICAR-Central Research Institute for Dryland Agriculture, CRIDA-NICRA Bulletin 2015/3, 22p.

3

Impact of Climate Change on Agriculture and Food Security

A.V.R Kesava Rao and Suhas P Wani

*ICRISAT Development Centre,
International Crops Research Institute for the
Semi-Arid Tropics (ICRISAT), Hyderabad.*

3.1 Introduction

Evidences over the past few decades show that significant changes in climate are taking place all over the world as a result of enhanced human activities in deforestation, emission of various greenhouse gases, and indiscriminate use of fossil fuels. Global atmospheric concentration of CO_2 has increased from pre-industrial level of 280 parts per million (ppm) to 400 ppm in 2014. Global projections indicate higher temperature of 1.5 to 4.5°C by the year 2050, as a result of enhanced greenhouse gases. Climate change predictions for India indicate that warming is likely to be above the global mean and fewer very cold days are very likely. Frequency of intense rainfall events and winds associated with tropical cyclones are likely to increase.

Various studies show that climate change in India is real and it is one of the major challenges faced by Indian Agriculture, more so in the semi-arid tropics (SAT) of the country. India ranks first among the countries that practice rainfed agriculture in terms of both extent and value of production. Rainfed agriculture is practiced under a wide variety of soil types, agro-climatic and rainfall conditions. Rainfed agriculture supports nearly 40% of India's estimated population of 1.21 billion in 2011 (Sharma 2011). The rainfed agro-ecologies cover about 60 per cent of the net sown area of 141 million ha and are widely distributed in the country (DOAC, 2011). Even after achieving the full irrigation potential, nearly 50% of the net cultivated area may remain dependent on rainfall. Changes in climate would affect agriculture directly through abiotic stresses and indirectly through biotic stresses. Climate change is seen as changes in temperature, increased variability in rainfall, enhanced carbon dioxide concentrations. Climate change is likely to make changes in the length of the rainfed crop-growing period. Rainfed agriculture in India plays a crucial role in ensuring food security for the larger and poorer segment of the population but often it coincides with a high incidence of poverty and malnutrition. Reduction in yields due to climate change is likely to be more prominent in rainfed agriculture and under limited water supply situations.

Crop yields in dryland areas of the country are quite low (1-1.5 t ha^{-1}) which are lower by two to five folds of the yields from researchers' managed plots (Bhatia et al., 2006). Current rainwater use efficiency in dryland agriculture varies between 35-45% and vast potential of rainfed agriculture could be unlocked by using available scientific technologies including improved cultivars.

3.2   Climate Change and Variability

Due to anthropogenic activities, a steady increase in atmospheric turbidity is observed in India. Indian annual mean (average of maximum and minimum), maximum and minimum temperatures showed significant warming trends of 0.51, 0.72 and 0.27°C 100 yr^{-1}, respectively, during the period 1901–2007 (Kothawale et al. 2010). However, accelerated warming was observed in the period 1971–2007, mainly due to intense warming in the recent decade 1998–2007.

Mean annual temperature of India in 2010 was +0.93°C above the 1961-1990 average and the India Meteorological Department (IMD) declared that 2010 was the warmest year on record since 1901. Mean temperature in the pre-monsoon season (March-May) was 1.8°C above normal during the year 2010.

At the country scale, no long-term trend in the southwest monsoon rainfall was observed, although an increasing trend in intense rainfall events was reported. Goswami et al. (2006) analysed gridded rainfall data for the period 1951-2000 and found significant rising trends in the frequency and the magnitude of extreme rain events, and a significant decreasing trend in the frequency of moderate events over central India during the monsoon seasons. The seasonal mean rainfall does not show a significant trend, because the contribution from increasing heavy events is offset by decreasing moderate events. They concluded that a substantial increase in hazards related to heavy rain is expected over central India in the future. Increased frequency and intensity of extreme weather events in the past 15 years were reported (Samra et al. 2003 and 2006).

A study carried out by ICRISAT under the National Initiative on Climate Resilient Agriculture (NICRA) project described a net reduction in the dry sub-humid area (10.7 m ha) in the country, of which about 5.1 Million ha (47%) shifted towards the drier side and about 5.6 Million ha (53%) became wetter, comparing the periods 1971-1990 and 1991-2004 (Kesava Rao et al., 2013). Results for Madhya Pradesh have shown the largest increase in semi-arid area (about 3.82 Million ha) followed by Bihar (2.66 Million ha) and Uttar Pradesh (1.57 Million ha). Relatively little changes occurred in AP; semi-arid areas decreased by 0.24 Million ha, which were shifted to both towards drier side (0.13 Million Ha under arid type) and wetter side (0.11 Million Ha under dry sub-humid type). Results indicated that dryness and wetness are increasing in different parts of the country in the place of moderate climates existing earlier in these regions.

Weather-generating models are widely used for studying the climate change over longer periods. Reddy et al., (2014), have evaluated the LARS-WG model for southern Telangana region (Hayathnagar, Yacharam and Rajendranagar). A 30-year base weather data (1980–2010) was used to generate the long-term weather series from 2011 to 2060. The model predicted the maximum increase in average annual rainfall of 5.2% in 2030 and 9.5% in 2060 for Yacharam compared to Hayathnagar and Rajendranagar over the normal annual rainfall of the base period (1980–2010). In case of air temperature, the model predicted increase in maximum temperature in the range 1–1.5% and 2.5% for 2030 and 2060 respectively, for these locations whereas minimum temperature decreased in the range 3.7–10.2% and 6.3– 11.7% respectively, for 2030 and 2060.

In the northern districts of Karnataka, annual rainfall variability analysis based on hundred years' data (1901-2000) indicated a periodicity of 13-17 years' cycle. A general decrease of rainfall in September and large increase in October over northern Karnataka during the 20th century was seen. Corresponding decrease in maximum temperature during October was noticed. Increased temperatures in November and December are observed, indicating availability of higher thermal energy for better vegetative growth during November while greater thermal stress during flowering period in December for post-rainy sorghum crop (Venkatesh *et al.*, 2008).

Using a high spatial resolution (0.25° × 0.25°, latitude × longitude) climate data for the period 1901-2013, pixel-wise water balances and climate indices were computed based on the revised water budgeting approach of Thronthwaite and Mather (1955). Climates for each year were classified based on the annual moisture index as per classification of Thornthwaite and Mather (1955).

The period 1901-1990 is considered as the base period or period 1 and 1991-2013 is considered as period 2. Average climates classified into six types for both the periods 1 and 2. Considerable changes in climates are observed between the two periods, 1901-90 and 1991-2013 (Table 3.1). Arid area has decreased by about 0.574 m ha and much of it has shifted to the Semi-Arid climates. Interesting feature is that there is an eastward shift in the arid areas in the districts of Belgaum, Bagalkot, Koppal and Raichur. Areas under Semi-Arid and Perhumid climates have increased by about 0.334 and 0.316 m ha. Dry Sub-humid areas increased by about 0.12 m ha and Humid areas are reduced by about 0.161 m ha. Very little decrease of 0.035 m ha was observed in the Moist Sub-humid type of climate. Overall it is seen that Per humid, Dry Sub-humid, Moist Sub-humid and Semi-Arid areas have increased by about 27, 13, 8 and 3 per cent respectively, compared to their normal 90-year climate as seen in the period 1 (1901-1990). Arid and Humid climate areas have decreased by about 15 and 4 per cent respectively.

Table 3.1 Area changes under different climates in Karnataka

Area in million ha

Period	Arid	Semi-Arid	Dry Sub-humid	Moist Sub-humid	Humid	Per Humid	Total
1901-1990	3.775	11.688	0.929	0.452	1.149	1.186	19.179
1991-2013	3.201	12.022	1.049	0.417	0.988	1.502	19.179
Difference	-0.574	0.334	0.12	-0.035	-0.161	0.316	0
% Change	-15	3	13	8	-4	27	0

3.3 Climate Change Impacts

Due to global warming, length of the growing period (LGP) is likely to increase, however due to increase in day and night temperatures, physiological development is accelerated resulting in hastened maturation and reduced yields. Increased night time respiration may also reduce potential yields. With global climate change, rainfall variability is expected to further increase. When decrease in rainfall coupled with higher atmospheric requirements due to elevated temperatures, the LGP is likely to shorten. At Nemmikal watershed in the Nalgonda district of Telangana, the LGP has decreased by about 15 days since 1978 and the climate has shifted to more aridity from semi-arid (Wani et al., 2012). Shift in the length of growing period, if not understood by the farmers, generally results in more crop failures due to late season drought (Fig. 3.1). Present popular varieties of maize and pigeonpea are likely to produce lower yields more often in future.

In Karnataka, Eastern Dry agroclimatic zone consists of Bangalore and Kolar districts and parts of Tumkur district, which is also known as the Tank-fed region. Rajegowda *et al.*, 2000 have shown that there is a predominant shift in the initiation and termination of rainfall to supply adequate moisture for crop growing period. This shift has been observed after 1990 and their mean monthly values also have changed. Before 1990, the annual rainfall ranged from 619 to 1119 mm with a mean of 869 mm.

After 1990, the annual rainfall ranged between 611 and 1311 mm with a mean of 1011 mm. During the first period, on an average, the peaks were observed during May, July and September while during the second period, the peaks were observed during May, August and October. There is a perceptible shift in rainfall pattern from July to August and also from September to October in this agroclimatic zone.

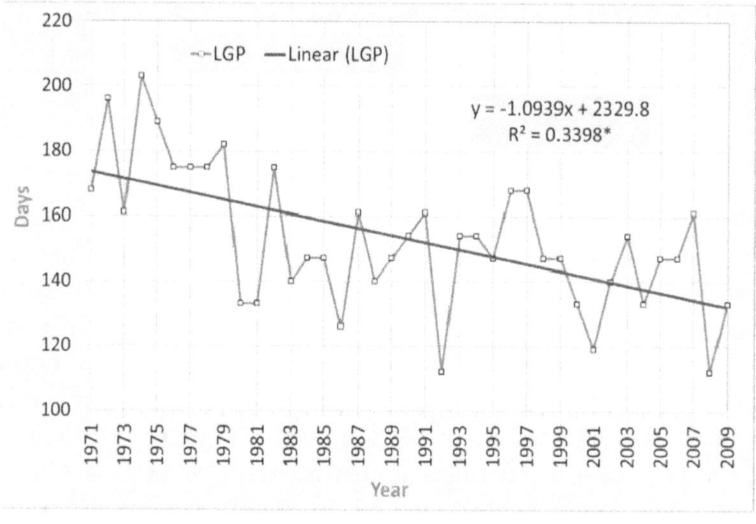

Fig. 3.1 Shift in Length of Growing Period at
Nemmikal, Nalgonda district

Crops sown during July rains would reach the grand-growth period viz., flowering to grain formation stage (long duration crops of about 115 days) during September which was receiving the highest rainfall till 1990, so that there was no moisture stress during the grand growth period. After 1990, as a result of reduction in July and September rains, the crops cannot be sown during July, though the land preparation could be done using June rains. Even with scanty rains, if the sowing is done during July, the crop would suffer from moisture stress due to the reduction in rainfall during September and also the crop grown would be caught in October rains causing considerable loss in the grain yield. This analysis revealed that in the Eastern Dry agroclimatic zone, sowing of crops (long duration variety crops of about 115 days) could be done during August preparing the land using June and July rains. In the years of early onset of southwest monsoon, sowing can be recommended during last week of July also. Crops sown during August would reach the grand growth period during October. As October receives higher rainfall the crop in its grand-growth period would not suffer for want of moisture and higher crop yields are expected. Crops sown beyond August may not be able to complete its life cycle as a result of inadequate moisture availability beyond 2nd fortnight of November (in the event of the intensity of northeast monsoon being low) as crop maturity coincides during this period. Under such circumstances, the short duration variety crops have to be preferred.

In a State like Karnataka having a spectrum of climates ranging from per humid type in the coastal and Malnad region to arid type in Bellary-Bijapur region, it is indicated that the Southwest monsoon rainfall is likely to be more uncertain with both increasing and decreasing trends in different parts of the state. Surface air temperature and diurnal temperature ranges are likely to increase along the high-ranges of the Western Ghats and under such conditions; there is a threat to thermo-sensitive crops like black pepper, cardamom, tea, coffee, cashew and other plantation crops. Frequent occurrence of droughts led to development of drought tolerant varieties to sustain agricultural production over the State. Unlike in other states, forest cover in Karnataka is increasing which is a positive sign. Karnataka State Action Plan on Climate Change was developed in the year 2012 which discusses climate trends, projected vulnerabilities along with adaptation and mitigation priorities for various sectors (EMPRI, 2012).

3.4 Climate Change Impacts on Crops

Rise in the mean temperature above a threshold level will cause a reduction in agricultural yields. A change in the minimum temperature is more crucial than a change in the maximum temperature. Grain yield of rice, for example, declined by 10% for each 1 °C increase in the growing season minimum temperature above 32 °C (Pathak *et al.*, 2003). Climate change impact on the productivity of rice in Punjab (India) has shown that with all other climatic variables remaining constant, temperature increases of 1 °C, 2 °C and 3 °C, would reduce the rice grain yields by 5.4%, 7.4% and 25.1%, respectively (Aggarwal *et al.*, 2009).

Field experiments and lab analyses were conducted at IARI, New Delhi in 2005, with five high-yielding rice varieties including aromatic and non-aromatic types, exposed to twelve different diurnal temperature (day/night) and radiation regimes to ascertain the impact of diurnal temperature and radiation changes on yield and yield components of aromatic and non-aromatic rice varieties in the field conditions and to document their effect on the grain and seed quality. Salient results indicate that the grain yield of all the five varieties was most significantly influenced by MNT (P < 0.001), followed by radiation (P < 0.001), explaining 87% and 77% of the yield variation respectively (Anand et al., 2015). Highest yields were recorded around a very narrow optimum temperature of 23°C to 24°C, with subsequent increase in temperature even by 1°C or 2°C, significantly reducing the grain yield.

General Circulation Models (GCMs) representing physical processes in the atmosphere, ocean, cryosphere and land surface are the most advanced tools currently available for simulating the response of the global climate system to increasing concentration of greenhouse gases. The GCMs depict the climate using a three-dimensional grid over the globe, typically having a horizontal resolution between 250 km and 600 km (approx. 2.5° × 2.5°). Their resolution is thus quite coarse relative to the scale of exposure units in most impact assessments. Regional climate models (RCMs) are run with the inputs from GCMs as well as from local topographical information. These models are of finer resolution, i.e., about 25 km × 25 km or less and thus are vital for regional impact assessments.

Simulation models are strong tools which provide opportunity to use various climate change scenarios in combination with different management parameters for analysing the regional impacts (Naresh Kumar and Aggarwal, 2009). Crop growth simulation models are used to quantify the impacts of elevated temperature, increased CO_2 level, change in rainfall, etc. individually or in combination. These analyses do provide vital information with reference to 'fixed changes in weather factors'. Another approach is to use the climate scenarios for impact assessments, wherein the global or regional climate model outputs of climate data are used as input for the crop simulation models to quantify the impacts. In fact, the research on climate change uses a cascade of simulation models (Naresh Kumar *et al.*, 2011).

3.4.1 Climate Change Impacts on Groundnut

Groundnut is the major oil seed crop in India and plays a major role in bridging the vegetable oil deficit in the country. Major groundnut growing states in India are Gujarat, Andhra Pradesh, Tamil Nadu, Karnataka and Maharashtra and groundnut is grown mostly under rainfed conditions and sensitive to moisture stress at different phenological stages. Uncertainty of rains is one of the major constraints for rainfed groundnut production. A year receiving low or below normal rainfall need not result in low productivity and similarly a year receiving high rainfall may not produce higher crop yields; well distributed rainfall is the key for sustainable yields. Assessment of changes in duration and frequency of dry spells and wet spells in the groundnut crop-growing period due to climate change is important for planning and implementing suitable water management practices. Genetic coefficients for the groundnut variety ICGV 91114 were derived from the field experimental data. Daily observed weather data of ICRISAT Patancheru for 30 years (1985-2014) and DSSAT PnutGro model were used. Impacts of both 10-day and 15-day period water stress on groundnut were studied using PnutGRO model (Kesava Rao et al., 2015). Based on the future climate data sets, impacts of climate change on groundnut productivity at Patancheru were assessed.

Results indicated that a 15-day water stress period during 40-60 days after sowing (Fig. 3.2) would reduce yields significantly. If the water stress in this period is not properly managed, groundnut yields could be reduced by about 33 per cent of the rainfed potential yields.

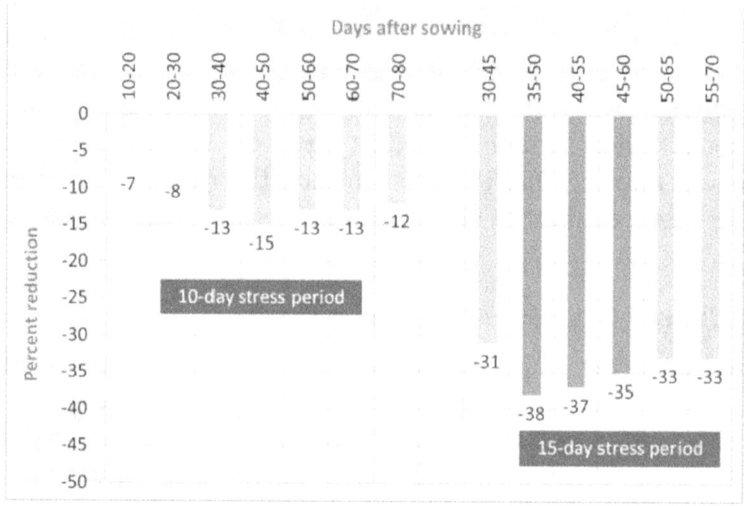

Fig. 3.2 Impact of water stress on groundnut pod yield

Current conditions (interpolations of observed data, representative of 1950-2000) and downscaled GCM data from Coupled Model Intercomparison Project Phase 5 (CMIP5) for three models (HadGEM2-ES, GFDL-CM3 and CNRM-CM5) were downloaded from the WorldClim – Global Climate Data portal for the whole world and from these datasets, representative data were extracted for the study location i.e., ICRISAT, Patancheru. These climate projections are for RCP 8.5 and for the year 2050 (2041 to 2060). Simulations with projected climate data (HadGEM2-ES, GFDL-CM3 and CNRM-CM5) indicated that groundnut pod yield would reduce by 9 to 13 per cent (Table 3.2).

Table 3.2 Impacts of projected climate on groundnut yields at ICRISAT

CC Scenario	Pod / Seed yield (kg ha^{-1})	Change in Pod / Seed yield (%)	Total Dry Matter production (kg ha^{-1})
Groundnut on Alfisols			
Current	2000	-	5430
HadGEM2-ES	1820	-9	5410
GFDL-CM3	1830	-9	5350
CNRM-CM5	1750	-13	5250

Simulations based on observed data have shown that if the water stress in this period is not properly managed, groundnut yields could be reduced by about 33 per cent of the rainfed potential yields.

3.4.2 Climate Change Impacts on Pigeonpea

Pigeonpea is the second most important pulse crop after chickpea in India. In 2010-11, it is cultivated in about 4.37 M ha (17% of the total area under pulses in the country) and contributs to about 16% to the total pulses production with an average productivity of 0.66 t ha⁻¹. In Karnataka, Gulbarga area is known as "Pulse Bowl of Karnataka". Pigeonpea area in Gulbarga has increased by three-folds from about 0.14 M ha in 1970 to 0.43 M ha in 2007. There has been a sharp and steady increase in area under pigeonpea since 1995. Average pigeonpea productivity, however, is low at about 0.42 t ha⁻¹. Pigeonpea productivity in Gulbarga is affected by large variations in rainfall amount and distribution, increased temperatures and depleting soil productivity. Genetic coefficients for the popular and promising pigeonpea variety TS-3R were estimated based on field experiments at ICRISAT, Patancheru.

Simulated pigeonpea grain yield and total biomass at Gulbarga were 2057 and 8708 kg ha⁻¹, respectively under baseline (present) climate. Increase in temperature by 1 and 2 °C could decrease grain yield by 9 and 16%, respectively (Table 3). Similarly, total biomass decreased by 5 and 9% with increase in the temperature by 1 and 2 °C. Decrease in rainfall by 10% coupled with increase in temperatures by 1 and 2 °C could further reduce grain yields by 5 and 4% making the total reduction at 14 and 20%. The situation could further worsen with reduction in rainfall by 20%, making the loss of grain yields by 21 and 28% with increase in temperature by 1 and 2 °C, respectively. Increased rainfall scenarios could benefit the crop to some extent, particularly in the low rainfall years, but net effect still remained negative.

Increased temperature could shorten the crop duration. Days to flowering shortened by 2 and 4 and the total crop duration by 5 and 9 days with increase in temperature by 1 and 2 °C, respectively. Increase in temperature causes more transpiration per day which results in water stress during the dry periods. Water balance outputs have shown that decrease in rainfall by 10 and 20% resulted in less plant water use by 18 and 45 mm, respectively with increase in temperature by 2 °C. Increments in rainfall by 10 and 20% are likely to result in more rainfall only for those days with rainfall and will not affect non-rainy days. Thus, additional rainfall has contributed more towards runoff and drainage than evapotranspiration. Simulated water use efficiency of pigeonpea reduced from 7.2 kg ha^{-1} mm^{-1} in the baseline by 6.6 and 6.0 kg ha^{-1} mm^{-1} with temperature increase of 1 and 2 °C, respectively. Better water and nutrient management approach is the key and Integrated Watershed Management plays a major role in sustaining pigeonpea productivity under future climate scenarios. Adoption of varieties tolerant to high temperature could also play a major role for sustainable pigeonpea yields. Water stress during the end of season could be avoided by sowing the short and extra-short duration varieties. Breeding of varieties which can put extra root mass is required for sustainable pigeonpea production.

Table 3.3 Effect of projected climate on phenology and productivity of pigeonpea cv. TS-3R

Climate scenario	Days to flower	Days to maturity	Total biomass (kg ha^{-1})	Grain yield (kg ha^{-1})	Change in yield (%)
Present (P)	103	157	8708	2057	0
P+1°C	101	151	8286	1875	-9
P+1°C-10%RF	99	150	7798	1771	-14
P+1°C-20%RF	99	150	7090	1615	-21
P+1°C+10%RF	101	151	8659	1961	-5
P+1°C+20%RF	101	152	8866	2005	-3
P+2°C	99	148	7943	1734	-16
P+2°C-10%RF	98	147	7465	1636	-20
P+2°C-20%RF	98	147	6763	1486	-28
P+2°C+10%RF	100	149	8302	1809	-12
P+2°C+20%RF	99	148	8525	1854	-10

Climate change affects dynamics and interaction among species and it will affect and change the pattern of pest damage and pest control strategies. Increase in temperature may increase the need for application of pesticides and may reduce pesticide effectiveness and increase residues. Rao *et al.* (2009) have conducted feeding trials with two foliage feeding insect species, *Achaea janata* and *Spodoptera litura* using foliage of castor plants grown under four concentrations of CO_2, viz. 700 ppm CO_2 inside open top chamber (OTC), 550 ppm CO_2 inside OTC, ambient CO_2 (350 ppm) inside OTC and ambient CO_2 in the open. Biochemical analysis of the foliage revealed that plants grown under the elevated CO_2 levels had lower N-content, and higher C-content, C/N ratio and polyphenols. Compared to the larvae fed on the ambient CO_2 foliage, the larvae fed on 700 ppm and 550 ppm CO_2 foliage exhibited higher consumption. The 700 ppm and 550 ppm CO_2 foliage was more digestible with higher values of approximate digestibility. The relative consumption rate of larvae increased, whereas the efficiency parameters, viz. efficiency of conversion of ingested food (ECI), efficiency of conversion of digested food (ECD) and relative growth rate (RGR) decreased in the case of larvae grown on 700 ppm and 550 ppm CO_2 foliage. The consumption and weight gain of the larvae were negatively and significantly influenced by the leaf nitrogen, which was found to be the most important factor affecting consumption and growth of larvae.

Using the 'Rice FACE' facility in northern Japan, Kobayashi *et al.* (2006) studied the effect of 200–280 ppm above-ambient CO_2 on rice blast and sheath blight disease for three seasons. Severity of leaf blast (*Magnaportheoryzae*) was consistently higher at the elevated CO_2 levels in all the three years assessed at two different stages of rice growth.

3.5   Food Grain Production Trends

From a net importer of food in 1950s, India has transformed itself in the production of food grains (mainly rice, wheat, coarse cereals and pulses) during the last few decades. From a mere 50 million tons (mt) of annual food grain production in 1950s, India in 2012-

13 has produced 257 mt of food grains, mainly attributed to the significant jump in rice and wheat output. The average growth rate of food grains production from 1950 to 2011 was 3.2% per annum. Overall, wheat was the best performer, with production increasing from mere 6.6 mt in 1950-51 to 90 mt during 2011-12, a huge jump. Wheat was followed by rice, which had a production increase from 20 mt to 102 mt. Total food grains production in India, Andhra Pradesh, Karnataka and Telangana are shown in the figures 3.3, 3.4, 3.5 and 3.6. It is seen that both India as a whole and Andhra Pradesh show consistent increase in productivity, it appears that in Karnataka it varied from about 1 to 1.8 tonne per hectare. In Telangana, the productivity has shown great improvements in the recent years even touching 3 tonne per hectare.

Cereals productivity in Northern Karnataka is very low; since large areas are under cultivation, there is great potential for improvement. Pulses productivity levels are far below their potential. Except the three northern districts (Bidar, Bijapur and Gulbarga), cultivation of pulses is very low. Crop rotations with pulses will improve overall food grain productivity in Karnataka.

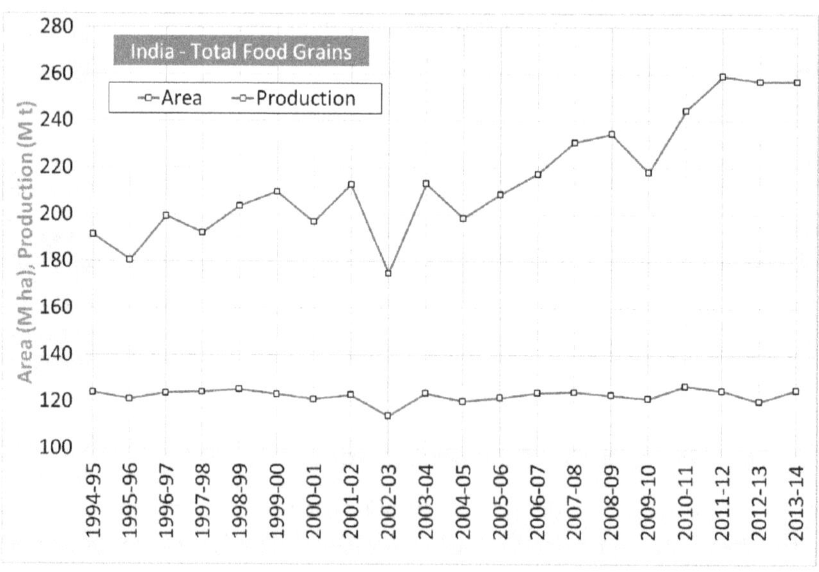

Fig. 3.3 Variability in area and production of total food grains in India

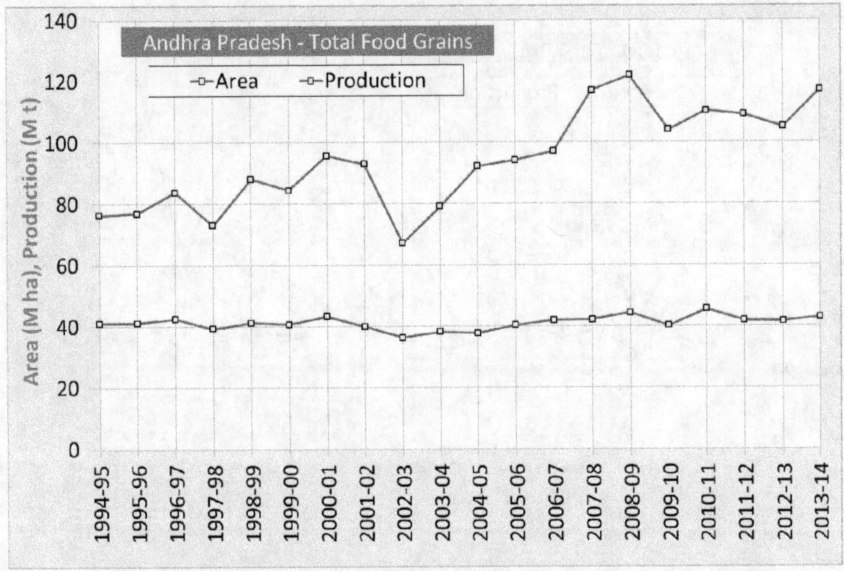

Fig. 3.4 Variability in area and production of total food grains in Andhra Pradesh

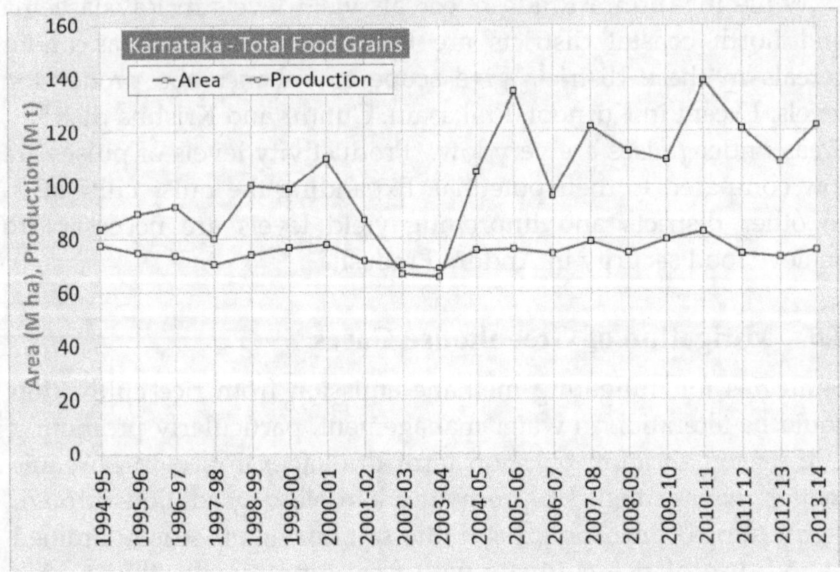

Fig. 3.5 Variability in area and production of total food grains in Karnataka

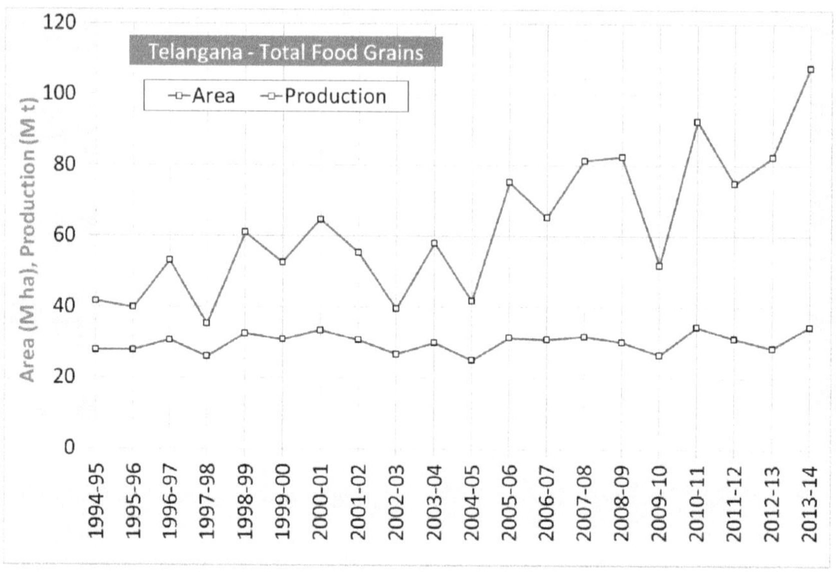

Fig. 3.6 Variability in area and production of total food grains in Telangana

While in Andhra Pradesh, cereals yield levels in Rayalaseema and north coastal districts are low. Considerable areas under cereals in these districts give scope to enhance the production levels. Except in Kurnool, Prakasam, Guntur and Krishna districts, areas under pulses are very low. Productivity levels of pulses are low compared to their potential. Expanding the pulse cultivation in other districts and improving yield levels are necessary to achieve food security in Andhra Pradesh.

3.6   Mitigation of Greenhouse Gases

Strategies for mitigating methane emission from rice cultivation could be alteration in water management, particularly promoting mid-season aeration by short-term drainage; improving organic matter management by promoting aerobic degradation through composting or incorporating it into soil during off-season drained period; use of rice cultivars with few unproductive tillers, high root oxidative activity and high harvest index; and application of fermented manures like biogas slurry in place of unfermented farmyard manure (Pathak and Wassmann, 2007). Methane emission from ruminants can be reduced by altering the feed composition, either to reduce the percentage which is converted into methane or to improve the milk and meat yield. The most efficient management practice to reduce nitrous oxide emission is site-specific, efficient nutrient management (Pathak, 2010). The emission could also be reduced by nitrification inhibitors such as nitrapyrin and dicyandiamide (DCD).

3.7 Leaf Colour Chart for Site-Specific Nitrogen Management

Leaf Colour Chart (LCC) is an easy-to-use and inexpensive tool for determining nitrogen status in plants. Use of the LCC promotes timely and efficient use of N fertilizer in rice and wheat to save costly fertilizer and minimize the fertilizer related pollution of surface water and groundwater. It is a promising eco-friendly and inexpensive tool in the hands of the farmers.

3.8 Climate Projections (IPCC AR5)

Approved summary for the Working Group I of the IPCC indicates that warming of the climate system is unequivocal, and since the 1950s, many of the observed changes are unprecedented over decades to millennia. Total radiative forcing is positive, and has led to an uptake of energy by the climate system. Continued emissions of greenhouse gases will cause further warming and changes in all components of the climate system. Global surface temperature change for the end of the 21st century is *likely* to exceed 1.5°C relative to 1850 to 1900 for all RCP scenarios except RCP2.6. It is *likely* to exceed 2°C for RCP6.0 and RCP8.5, and *more likely than not* to exceed 2°C for RCP4.5. Changes in the global water cycle in response to the warming over the 21st century will not be uniform. The contrast in precipitation between wet and dry regions and between wet and dry seasons will increase, although there may be regional exceptions.

3.9 Adaptation Strategies for Climate Resilience

There is a need to understand the shifts in climate at different agro-eco regions in SAT India in terms of anticipated shifts in the crop growing periods and water availability and to use the results of agroclimatic trend analysis to device suitable adaptation strategies. Increased dependence on groundwater irrigation and indiscriminate usage of chemical fertilizers for increasing crop productivity will not be sustainable in the future. Under the future likely climate change scenario, this situation would worsen jeopardising the long-term perspective of livelihoods of the great majority of the rural population.

Resilience to climate change depends on identifying climate smart crops and management practices and degree of awareness of community. Knowledge dissemination to farmer about climate change is critical. The present weakest link of knowledge dissemination needs to be strengthened. Innovative participatory delivery systems using ICT mediated climate early warning systems and weekly agromet advisories would empower farmers of Karnataka to take smart and swift decisions and make them resilient and food secure. There is a need for scaling up the integrated climate smart agricultural practices for bringing in resilience. Social safety nets along with adaptation strategies are needed. Another major activity required under crop insurance scheme is to identify the weather based indices for various rainfed crops and varieties.

Intercropping with grain legumes is one of the key strategies to improve productivity and sustainability of rainfed agriculture. Productive intercropping options identified to intensify and diversify rainfed cropping systems are:

- Groundnut with maize
- Pigeonpea with maize
- Pigeonpea with soybean

Some of the other initiatives are ridge planting systems; seed treatment; Integrated Pest Management (IPM); adoption of improved crop varieties and production technologies; promoting community-based seed production groups and market linkages. Farmers need to be encouraged to practice seed treatment with Trichodermaspp and fungicides for managing seedling diseases and IPM options for controlling pod borer in chickpea and pigeonpea.

3.9.1 Polythene-lined Water Storage Ponds in Karnataka

About 35,000 farmers across the Karnataka state's 175 taluks are implementing the pilot programme by setting up polythene-lined water storage ponds in their fields to prevent water seepage and store run-off rainwater. The intent is to ensure that the dryland farmer has at least some water to ensure that their entire crop is not ruined, either through excess rain or due to drought, said by Karnataka's minister of state for agriculture, Krishna Byre Gowda. All of them have individual or jointly-owned diesel pumps that will provide micro irrigation (drip irrigation) from the poly ponds for their fields. It is being piloted for the kharif season 2015 that began in April and will go on until September. Another 1,750 farmers have opted for more high-end poly houses - poly roofs over a section of their farms - to grow exotic vegetables under shade, which will ensure that they won't fall into debt under any circumstances, be it excess or deficit rain.

3.9.2 Poly Mulching Technology for Better Quality Tomato

Summer season is a time of worry for most farmers across the country since water becomes an important, and much sought after commodity. Though water harvesting and conservation are being encouraged by the government the number of farmers adopting it is still quite negligible in the country. Small farmers cannot afford to dig a small pond to collect rainwater since it eats away into their cropping area. For such growers, the Krishi Vigyan Kendra (KVK) under the Indian Institute of Horticultural Research (IIHR, Bangalore) at Hirehalli, Tumkur, Karnataka, initiated demonstrations to popularize the practice of poly mulching technology in the region. The sheets are laid on the field by a machine on top of the furrows and seedlings are planted in small holes made on the sheets. Plastic sheets have been found to conserve soil moisture because the water that gets evaporated from the soil in the open will condense on the lower part of the sheet as small droplets and falls back into the soil.

A small farmer, Ms Saroja, from Deverayanapatna village in Tumkur taluk, with having two acres of land was encouraged to grow the tomato variety ArkaSamrat released by IIHR under this technology. The normal duration of this variety is 135-140 days only but due to the impact of polythene mulching, the crop period extended to 10-15 days more. The tomato seedlings were grown on raised beds with poly mulch film laid with drip irrigation. A package of practices like mulching was suggested which minimised the incidences of pests and viral diseases as reported by The Hindu on 30 July 2014.

The fruits obtained are of better quality and colour, which fetch better price in the market. The farmer harvested nearly 32 tonnes from an acre in 150 days and sold them at Rs.10 per kg in the local market. She earned a gross profit of Rs. 3.25 lakhs in 150 days. Total cost of cultivation was Rs. 60,000 per acre and the farmer earned a net profit of Rs. 2.65 lakhs in five months.

3.10 ICRISAT's Hypothesis of Hope to Address Climate Variability and Change

ICRISAT's research findings showed that Integrated Genetic and Natural Resources Management (IGNRM) through participatory watershed management is the key for improving rural livelihoods in the SAT (Wani *et al.*, 2002, 2003 and 2011). Even under a climate change regime, crop yield gaps can still be significantly narrowed down with improved management practices and using Germplasm adapted for warmer temperatures (Wani *et al.*, 2003, 2009 and Cooper et al., 2009). Some of the climate resilient crops are short-duration chickpea cultivars ICC 96029 (Super early), ICCV 2 (Extra-early) and KAK 2 (Early maturing); wilt resistant pigeonpea hybrid (ICPH 2671) with a potential to give 80% higher yields than traditional varieties and short-duration groundnut cultivar ICGV 91114 that escapes terminal drought.

Integrated Watershed Management comprises improvement of land and water management, integrated nutrient management including application of micronutrients, improved varieties and integrated pest and disease management for substantial productivity gains and economic returns by farmers (Wani *et al.*, 2003). The goal of watershed management is to improve livelihood security by mitigating the negative effects of climatic variability while protecting or enhancing the sustainability of the environment and the agricultural resource base. Greater resilience of crop income in Kothapally (Andhra Pradesh) during the drought year 2002 was indeed due to watershed interventions. While the share of crops in household income declined from 44% to 12% in the non-watershed project villages, crop income remained largely unchanged from 36% to 37% in the watershed village (Wani et al., 2009).

Agroclimatic analysis coupled with crop-simulation models, and better seasonal and medium duration weather forecasts, help build resilience to climate variability/ change in watersheds. This means that high yields are still possible under variable climate if farmers combine improved practices with climate-adapted crop varieties. Hence, the challenge today is to encourage rainfed farmers to adopt these improved options to increase productivity and profitability of their crops, and to become more resilient under climate variability. Farmers need to be encouraged to enhance soil quality and fertility through composting of organic wastes; and to promote cultivation of Leucaena, *Hardwickiabianta* and Glyricidia on farm bunds. Governments may also consider promoting and incentivizing the soil and water conservation measures taken by farmers. An improved agromet advisory service at the local level along with associated weather insurance packages is a sure way to enhance the resilience of poor farmers in the context of climate change. Policy interventions are needed to mitigate the climate change effects and Governments have to be proactive in developing adaptation strategies for those sectors like agriculture, water resources, forestry and biodiversity which are highly exposed to the future climate changes and have a significant impact on livelihoods.

Enhancing resilience through climate smart agriculture

ICRISAT's findings show that even under a climate change regime, crop yield gaps can still be significantly narrowed down with improved management practices and using Germplasm adapted for warmer temperatures. As an adaptation strategy to climate variability/change and in rainfed areas, ICRISAT has recommended the cultivation of climate resilient crops, such as short-duration chickpea cultivars, wilt resistant pigeonpea hybrids and short-duration groundnut cultivar ICGV 91114 that escapes terminal drought. Chickpea is traditionally grown in the relatively colder northern India during the dry winter season with temperatures of 20-30 °C and traditional chickpea varieties are of

---

Contd...

long duration (more than 120 days) and not very suitable for Andhra Pradesh. Ten years ago, only 160,000 hectares of chickpeas were grown in AP and the yield was only 600 kg ha^{-1}. Introduction of improved varieties of chickpea like JG11, *Swetha* (ICCV 2) and *Kranthi* (ICCC 37), which are resistant to higher temperature and wilt disease has changed the situation and the area under chickpea cultivation increased to 630,000 hectares and average yield increased from 600 to 1,400 kg ha^{-1}.

The community-based management of natural resources calls for new approaches (technical, institutional and social) which are knowledge-intensive and need strong capacity development measures for all the stakeholders including policy makers, researchers, development agents, and farmers. We need to connect the small and marginal farmers to the new knowledge and materials produced by the researchers.

ICRISAT's principle to improve the livelihoods of small-holder farmers even under future climate change scenario is built on the concept of Inclusive Market Oriented Development (IMDO), which is a Dynamic Development Pathway consisting of innovative environment, inclusive and market oriented.

In the years with aberrant weather, even the landless households face loss of employment opportunities and many of them may have to migrate to longer distances in search of employment opportunities. Production uncertainties also depress the prospects of trade and services. In short, the entire rural economy is highly vulnerable to the risks associated with climatic variability and possible climate change. Inputs of knowledge-intensive institutions like research, extension and development organizations become critical in enhancing the resilience of rural people to cope with the changing climatic patterns.

Weather based Crop Insurance Scheme (WBCIS) provides insurance protection against losses in crop yield resulting from adverse weather conditions. It provides pay-out against adverse rainfall incidence (both deficit & excess) during *Kharif* and adverse incidence in weather parameters like frost, heat, relative humidity, unseasonal rainfall etc. during *Rabi*. It is not yield guarantee insurance and is different from the National Agricultural Insurance Scheme (NAIS). One key advantage of the weather risk based crop insurance is that the pay-outs could be made faster, besides the fact that the insurance contract is more transparent and the transaction costs are lower. Under the changing climates, a key-challenge in designing insurance is to transfer risk and incentivize risk reduction through price incentives and risk management stipulations. Without this complementary risk reduction, more risks could become uninsurable in the future.

References

Aggarwal PK, Singh AK, Samra JS, Singh G, Gogoi AK, Rao GGSN and Ramakrishna YS (2009). Introduction. In *Global Climate Change and Indian Agriculture*, Ed: PK Aggarwal, ICAR, New Delhi, 1-5p.

Bhatia VS, Singh Piara, Wani SP, Kesava Rao AVR and Srinivas K. (2006). Yield Gap Analysis of Soybean, Groundnut, Pigeonpea and Chickpea in India using Simulation Modeling. Global Theme on Agroecosystems Report no. 31. Patancheru 502 324, Andhra Pradesh, India: International Crops Research Institute for the Semi-Arid Tropics (ICRISAT). 156p.

Cooper P, Rao KPC, Singh P, Dimes J, Traore PS, Rao AVRK, Dixit P and Twomlow SJ. (2009). Farming with Current and Future Climate Risk: Advancing a "Hypothesis Of Hope" for Rainfed Agriculture in The Semi-Arid Tropics. Journal of Sat Agricultural Research 7. An open access journal published by ICRISAT, Patancheru.

DOAC. (2011). Annual Report 2010-11. Department of Agriculture and Cooperation, Ministry of Agriculture, Government of India.http://agricoop.nic.in/Annual report2010-11/AR.pdf

Environmental Management & Policy Research Institute (EMPRI) and The Energy and Resources Institute (TERI), (2012). Karnataka State Action Plan on Climate Change, 1st Assessment.

Goswami BN, Venugopal V, Sengupta D, Madhusoodanan MS and Xavier PK. (2006). Increasing Trend of Extreme Rain Events over India in a Warming Environment. *Science* 314: 1442–1445.

Kesava Rao AVR, Suhas P Wani, KK Singh, M Irshad Ahmed, K Srinivas, Snehal D Bairagi and O Ramadevi. (2013). Increased Arid and Semi-arid Areas in India with Associated Shifts during 1971-2004. Journal of Agrometeorology 15(1): 11-18.

Kesava Rao AVR, Suhas P Wani, K Srinivas, Pushparaj Singh, Snehal D Bairagi and O Ramadevi. (2013). Assessing Impacts of Projected Climate on Pigeonpea Crop at Gulbarga. Journal of Agrometeorology, 15 (Special Issue II): 32-37.

Kesava Rao AVR, Suhas P Wani and Srinivas K. (2015). Methodologies for Asessing Climate Change Impacts on Groundnut Productivity.

Proceedings of the International Conference on "Climate Change and Social-Ecological-Economical Interface-Building: Modelling Approach to Exploring Potential Adaptation Strategies for Bio-resource Conservation and Livelihood Development", ISEC, Bangalore.

Kobayashi T, Ishiguro K, Nakajima T, Kim HY, Okada M and Kobayashi K (2006). Effects of Elevated Atmospheric CO_2 Concentration on the Infection of Rice Blast and Sheath Blight. *Phytopathology*, 96: 425-31.

Kothawale DR, Munot AA and Krishna Kumar K. (2010). Surface Air Temperature Variability over India during 1901–2007, and its Association with ENSO. Climate Research 42: 89-104.

Nagarajan S, Jagadish SVK, Hariprasad AS, Tomar AK, Anand A, Madan Pal and Aggarwal PK (2010). Effect of Night Temperature and Radiation on Growth, Yield and Grain Quality of Aromatic and Non Aromatic Rice. *Agri. Ecosys. Environ.* 138: 274-281.

Naresh Kumar S and Aggarwal PK (2009). Role of Crop Simulation Models in Climate Change Research. In: *Photosynthetic Efficiency and Crop Productivity under Climate Change Scenario* Eds: M. P. Singh, P. Kumar, V. Jain, R. Pandey, R. Pandy and S. Khetrapal. Indian Agricultural Research Institute, New Delhi. 1-3p.

Naresh Kumar S, Aggarwal PK, Rani S, Jain S, Saxena R and Chauhan N (2011). Impact of climate change on crop productivity in Western Ghats, coastal and northeastern regions of India. *Curr. Sci.* 101(3): 33-42p.

Pathak H, Ladha JK, Aggarwal PK, Peng S, Das S, Yadvinder Singh, Bijay-Singh, Kamra SK, Mishra B, Sastri ASRAS, Aggarwal HP, Das DK and Gupta RK (2003). Climatic Potential and On-farm Yield trends of Rice and Wheat in the Indo-Gangetic Plains. *Field Crops Res.*, 80(3): 223-234p.

Pathak H and Wassmann R (2007). Introducing Greenhouse Gas Mitigation as a Development Objective in Rice-based Agriculture: I. Generation of Technical Coefficients. *Agril. Syst.*, 94: 807-825.

Pathak H (2010). Mitigating Greenhouse Gas and Nitrogen Loss with Improved Fertilizer Management in Rice: Quantification and Economic Assessment. *Nutr. Cycling Agroecosys.* 87: 443-454.

Prabhakar Pathak, Suhas P Wani and Raghavendra Rao Sudi. (2011). Long-term Effects of Management Systems on Crop Yield and Soil Physical Properties of Semi-arid Tropics of Vertisols. Agricultural Sciences, Vol.2, No.4, 435-442p. doi:10.4236/as.2011.24056.

Rajegowda MB, Muralidhara KS, Murali NM and Ashok kumar TN. (2000). Rainfall Shift and its Influence on Crop Sowing Period. *Journal of Agrometeorology* 2(1), 2002.

Rao, AVRK, Wani SP, Piara Singh, Rao GGSN, Rathore LS and Sreedevi TK. (2008). Agroclimatic Assessment of Watersheds for Crop Planning and Water Harvesting. Journal of Agrometeorology. Vol. 10, No. 1, 1-8p.

Rao MS, Srinivas K, Vanaja M, Rao GGSN, Venkateswarlu B and Ramakrishna, YS (2009). Host Plant (*Ricinuscommunis*Linn.) Mediated Effects of Elevated CO_2 on Growth Performance of Two Insect Folivores. *Curr. Sci.* 97(7): 1047-1054p.

Reddy KS, Kumar M, Maruthi V, Umesha B, Vijayalaxmi and Nageswar Rao CVK. (2014). Climate Change Analysis in Southern Telangana Region, Andhra Pradesh using LARS-WG model. Current Science, Vol. 107, No. 54.

Samra JS, Singh G and Ramakrishna YS. (2003). Cold Wave of 2002-03: Impact on Agriculture. NRM Division, ICAR, Krishi Anusandhan Bhawan-II, Pusa, New Delhi, India.49p.

Samra JS, Ramakrishna YS, Desai S, Subba Rao AVM, Rama Rao CAYVR, Rao GG SN, Victor US, Kumar Vijaya, Lawande KE, Srivastava KL and Krishna Prasad VSR. (2006). Impact of Excess Rains on Yield, Market Availability and Prices of Onion. Central Research Institute for Dryland Agriculture (ICAR). 52p.

Sharma KD. (2011). Rain-fed Agriculture could meet the Challenges of Food Security in India. Current Science 100 (11).

Suhas P Wani, William D Dar, Dileep K Guntuku, Kaushal K Garg and AVR Kesava Rao. Improved Livelihoods and Building Resilience in the Semi-arid Tropics: Science-led, Knowledge-based Watershed Management, Climate Exchange, The Global Framework for Climate Services, World Meteorological Organization, October 2012, Pages 69-71, ISBN 978-0-9568561-3-5. (http://www.wmo.int/pages/gfcs/tudor-rose/index.html).

Thornthwaite CW and Mather JR. (1955). The Water Balance. Publications in climatology. Vol. VIII. No.1. Drexel Institute of Technology, Laboratory of climatology, New Jersey.

Venkatesh H, Krishna Kumar K, Aski SG and Kulkarni SN. (2008). Impact of Climate Change on Agriculture over northern Karnataka. *In:* Climate Change and Agriculture over India. *Eds.* Prasada Rao GSLHV, Rao, GGSN, Rao, VUM and Ramakrishna YS.Published and released under AICRP on Agrometeorology in connection with Int. Symp. on Agromet.

And Food Security, held at CRIDA (ICAR), Hyderabad during 18-21 Feb 2008. 113-128p.

Wani SP, Rego TJ, Pathak P and Singh Piara. (2002). Integrated Watershed Management for Sustaining Natural Resources in the SAT. Pages 227-236 in Proceedings of International Conference on Hydrology and Watershed Management. 18-20 December 2002, Hyderabad, India. Hyderabad, India: Jawaharlal Nehru Technological University (JNTU).

Wani SP, Singh HP, Sreedevi TK, Pathak P, Rego TJ, Shiferaw B and Iyer Shailaja Rama. (2003). Farmer-Participatory Integrated Watershed Management: Adarsha watershed, Kothapally India, An Innovative and Upscalable Approach. A Case Study. Pages 123-147 *in* Research towards integrated natural resources management: Examples of research problems, approaches and partnerships in action in the CGIAR (Harwood RR and Kassam AH, eds.). Interim Science Council, Consultative Group on International Agricultural Research. Washington, DC, USA.

Wani SP, Sreedevi TK, Rockström J and Ramakrishna YS. (2009). Rainfed Agriculture - Past Trend and Future Prospects. Pages 1-35 *in* Rainfed agriculture: Unlocking the Potential. Comprehensive Assessment of Water Management in Agriculture Series (Wani SP, Rockström J and Oweis T, eds.). CAB International, Wallingford, UK.

Wani SP and Rockström J. (2011). Watershed Development as a Growth Engine for Sustainable Development of Dryland Areas. Pages 35-52 *in* Integrated Watershed Management in Rainfed Agriculture (Wani Suhas P, Rockström J and Sahrawat KL, eds.). CRC Press, The Netherlands.

4

Vulnerability of Indian Agriculture to Climate Change and its Assessment

V. K. Sehgal, H. Pathak and S. D. Singh

Centre for Environment Science and Climate Resilient Agriculture
Indian Agricultural Research Institute, (IARI) New Delhi.

4.1 Introduction

Agriculture is a major source of food, nutrition and livelihood security in India. It involves almost two-thirds of the workforce in employment and accounts for a significant share in India's gross domestic product (GDP). Several industries depend on agricultural production for their requirement of raw materials. Due to its close linkages with other economic sectors, growth in agricultural sector has a multiplier effect on the economy of the country. The Indian agriculture has made significant progress in recent years. However, currently it is facing the challenges of stagnating net cultivated area, deteriorating the quality of land and water resources, reducing per capita land availability and factor productivity and climate change. The problem is highly challenging because more than 80% of Indian farmers are marginal (cultivating up to 1 hectare land) and small (cultivating 1-2 hectares land) with poor coping capacity. The farms are diverse, heterogeneous and unorganized. Indian agriculture, with almost 60% of its net cultivated area as rainfed, is exposed to stresses resulting from climatic variability and climate change. India has the problem of ensuring food security for the projected most populous country in 2050 with one of the largest malnourished populations. Climate is the primary determinant of agricultural productivity. Over the past few decades, the man-induced changes in the environment have intensified the risk of climate-dependent crop production. The most imminent of the climatic changes is the increase in atmospheric temperature due to the rising levels of greenhouse gases in the atmosphere (IPCC 2007). It has been manifested in terms of frequent occurrence of events like droughts, floods, storms, melting of glaciers and rise in sea levels. The amount of rainfall and its distribution has become highly uncertain. These changes are already appearing on the horizon and causing serious threat to food security of the nation. In coming years, such uncertainties and threats are going to intensify widely. The global mean surface temperature is projected to rise by 1.8–4.0 ºC by 2100 as per the report of Inter-Governmental Panel on Climate Change (IPCC 2007).

Increased temperature, uneven rainfall, decrease in irrigation water and extreme weather events are the potential consequences of the rise in global mean surface temperature. This will have a direct impact on agriculture sector, non-agriculture sector and natural resources which are directly linked to national economy. The destruction of agriculture and infrastructure has been observed by climate variability (like droughts and floods) which has a negative impact on human health and livelihood security. The rural people are particularly vulnerable to climate variability and changes owing to their heavy dependence on agriculture for food and livelihood. For preparing people to face these challenges, decision-makers and policy planners need information on climate change. A close assessment of the vulnerability i.e., the degree to which agriculture is susceptible to the adverse effects of climate change, including climate variability and extremes is needed to allocate resources effectively and reduce the impacts. Indian agriculture is primarily dependant on weather and any variation in its pattern affects agricultural production. Some areas of the country are more vulnerable than the others depending on their adaptive capacity and socioeconomic status. To address climatic vulnerability, decision-makers need to prioritize their responses for different regions as the resources are limited. The decision-makers should plan climate adaptation strategies based on vulnerability assessment and mapping regions for vulnerability.

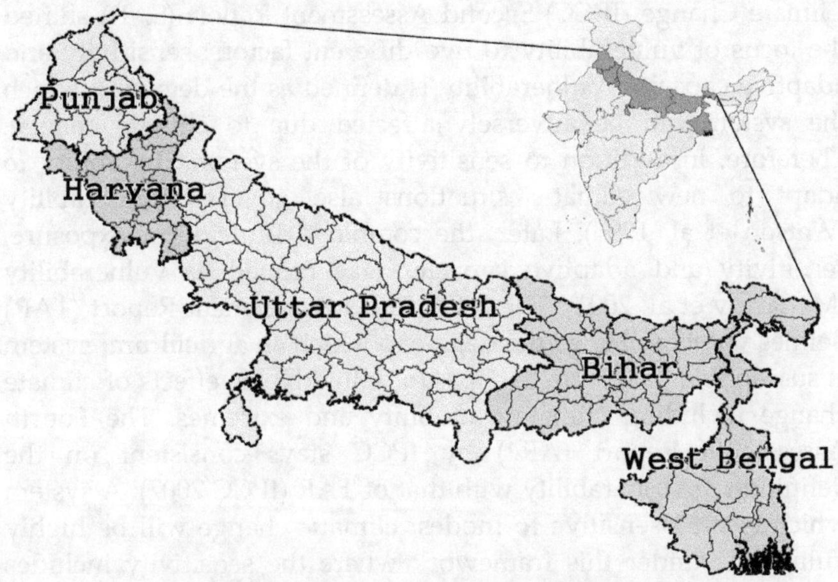

Fig. 4.1 The study area of the Indo-Gangetic Plains of India showing state and district boundaries

4.2 Vulnerability Assessment

Vulnerability to climate change is the degree to which an agricultural unit has the capacity to sustain the damage due to climate change, including climate variability and extremes. The process of identification, quantification and prioritization of vulnerability in a system is referred to as vulnerability assessment. The study of vulnerability or the degree, to which the people, environment or agriculture is affected, requires mainly three types of information:

1. Exposure, i.e., patterns of exposure to the occurrences of hazards such as droughts and floods;

2. Sensitivity, i.e., the degree to which the system can experience damages due to a particular event; and

3. Adaptive capacity, i.e. the capacity of a system to recover from disaster and hazards.

Vulnerability is mainly focussed on the internal coping and external exposure. However, the Intergovernmental Panel on

Climate Change (IPCC) Second Assessment Report (SAR) shifted the focus of vulnerability to two different factors: sensitivity and adaptive capacity. Vulnerability is defined as the degree to which the system can be adversely affected due to climate change. Therefore, in addition to sensitivity of the system, the ability to adapt to new climatic situations also governs vulnerability (Watson et al. 1996). Later, the combined function of exposure, sensitivity and adaptive capacity was termed as vulnerability (McCarthy et al. 2001). The IPCC Third Assessment Report (TAR) defines vulnerability as the degree to which an agricultural system is susceptible or unable to cope up with adverse effects of climate change including climate variability and extremes. The Fourth Assessment Report (AR4) by IPCC stays consistent on the definition of vulnerability with that of TAR (IPCC 2007). A system which is very sensitive to modest climatic change will be highly vulnerable under this framework where the sensitivity includes the potential for substantial harmful effects and for which the ability to adapt is severely constrained. Vulnerability to climate change is a multi-dimensional concept affected by various indicators and can be defined as a function of exposure, sensitivity and adaptive capacity of a particular system, i.e.,

$$Vits = f(Eits, Aits) \qquad \qquad .....(4.1)$$

where, Vits is the vulnerability of a system i to climate stimulus s in time t; Eits is the exposure of the system i to stimulus s in time t and Aits is the adaptive capacity of system i to deal with stimulus s in time t.

The IPCC gave a simpler way to define vulnerability (V) of a system as a function of exposure (E), sensitivity (S) and adaptive capacity (A), i.e.

$$Vulnerability = f\ (Exposure, Sensitivity, Adaptive\ capacity) \qquad .....(4.2)$$

$$Vulnerability = f\ (Potential\ Impact - Adaptive\ capacity) \qquad .....(4.3)$$

A higher adaptive capacity is associated with a lower vulnerability, while a higher impact is associated with a higher vulnerability. Given the above equation, vulnerability is defined as a function of a range of biophysical and socio-economic factors, aggregated into three components: exposure, sensitivity and adaptive capacity to climate variability and change. This study adopted the IPCC framework of vulnerability.

4.2.1 Exposure

The effects of climate change will be different at different locations. Some regions will be warmer than the others. Also, the precipitation patterns shift in different areas will be varying resulting in uneven distribution of rainfall. Some regions will see prolonged dry periods and some will experience both warm and intense rainfall. In correlations with the above statements, exposure relates to the degree of climate stress at a particular location. The exposure can also be determined by the long-term climatic changes or the variation in climate including the magnitude and frequency of hazards.

4.2.2 Sensitivity

The relative importance of the effects of climate change differs for different regions, groups and sectors in society. For example, highly intense rainfall may lead to devastating results in some region, whereas the same may not be of much harm in some other region. The degree to which a system is modified or affected by internal, external, or sometimes both disturbances is defined as sensitivity (Gallopin 2003). The measure that reflects the responsiveness of a system to climatic influences determines the degree to which a group will be affected by the environmental stress (SEI 2004).

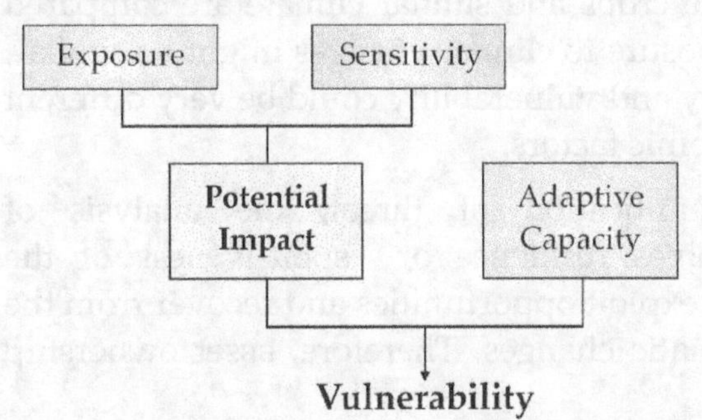

Fig. 4.2 Conceptual framework of assessing the vulnerability of agriculture to climate change (*Source:* IPCC 2007)

4.3 Adaptive Capacity

Depending upon sensitivity and exposure, the extent of response to the effects of climate change differs across regions. For example, frequent droughts can be addressed by some farmers by using appropriate irrigation technology, whereas other farmers may not be able to afford such technology or may lack the skills to operate it. Therefore, the ability to adapt to certain changes in condition is very important to determine the vulnerability of a system towards the change. Adaptability, coping ability, stability, management capacity, flexibility, robustness and resilience, all together form the ability of a system to adapt to the changes effectively. Therefore, 'Adaptive capacity' is a significant factor in characterizing vulnerability. Adaptive capacity is also defined as the potential or ability of a system, region or community to adjust to the effects or impacts of climate change (IPCC, 2001). Different countries, communities, social groups, individuals and times have different capacities to adapt. The adaptive capacity of a system or society is to deal with the changes in conditions to modify its own characteristics and behaviour. The increase in literacy levels enhances the capability of people to access information and cope up with adversities, resulting in reduced vulnerability. The farms with larger agricultural income, land area, farm value assets and latest technology are able to prepare and respond better as compared to the farms with lower technology. Also, the farms with traditional technologies are assumed to be less economically diversified and more vulnerable to climatic events. The availability of facilities like electricity, education, health care, etc., determines the state of poverty in a region. When two different agricultural regions having the same crops and similar climate are compared with each other, the exposure to climate changes might be similar, but the adaptive capacity and vulnerability could be very different based on the socio-economic factors.

In addition to identification of threat, the analysis of vulnerability also involves resilience or sponsiveness of the system and its ability to exploit opportunities and recover from the environmental and climatic changes. Therefore, asset ownership goes hand in hand with vulnerability. The people having more assets are less vulnerable to climate change.

4.4   Multi-criteria Decision-making (MCDM) Technique

The data from various domains and sources like meteorology, soil science, social science, etc., are processed to form an integrated approach to vulnerability. The various domains and streams coined as exposure, sensitivity, and adaptive capacity are grouped accordingly. The factors and attributes involved in each of these groups are not equally important for vulnerability. Some of the criteria contribute heavily towards it, whereas the others may have minimal importance for it. These criteria are grouped and organized in different hierarchies to address the relative degree of importance towards vulnerability. Multi-criteria evaluation techniques can be used to determine the suitability by evaluating the relative importance of these parameters. These techniques are well equipped to make decisions for agricultural applications with their ability to provide rational, objective and non-biased approach.

4.5   Framework for Decision-making

The general framework for assessing vulnerability by using decision making process in the following three phases:

Intelligence phase: This is the initial phase of identifying the problems for decision-making. The situation is analyzed for the problem and various prospects. As this phase also involves evaluation of the criteria according to the defined problems, this phase is also called as problem formulation phase.

Design phase: When the problems are defined, we need to understand them and generate alternatives, select crieria, establish relationship among them and assess the importance of given criteria. The design is basically formulated on multi-criteria decision-making methods.

Decision choice: Once the design phase is over, we need to evaluate the options and make decisions to deal with problems using multi-criteria decision rules.

4.6   Analytical Hierarchy Process

There are various types of comparisons between two elements; however, when the comparison is about the relative importance, the Analytical Hierarchy Process (AHP) is the best method for decision-making. As the name suggests, this method uses the analytical approach of decomposing the complex problems into its hierarchies and simpler groups. The empirical studies suggest that more than three criteria can not be compared at a given time. Therefore, the decomposition of complex decision-making processes into a hierarchical organization of criteria is helpful for decision-making. Another benefit of using a hierarchical organization is that the structure helps to maintain consistency amongst comparisons along with incorporation of decisions and expert knowledge from various domains. The decisions about the relative importance of criterion 'A' over some other criteria 'B', 'C', 'D', and 'E' are used to prepare a pair-wise comparison matrix which is the basic input to AHP. Ratio matrix of criteria is produced along with the relative weight of each criterion using pair-wise comparisons. The relative importance of a particular criterion over other criteria in consideration is termed as the 'weight' of that criterion. The criterion is more important if the weight is higher (Malczewski 1999). Eigen-value of the ratio-matrix is used to determine the weights by normalizing the Eigen-vectors associated with a criterion.

4.7   Past Studies on Vulnerability of Agriculture to Climate Change

Several vulnerability assessment studies have been done in different fields including climate, agricultural sciences, social sciences, geography and environmental sciences. Some analysts have used theoretical perspectives to define the nature of vulnerability, while the others have developed some quantitative measures for vulnerability. Vulnerability assessments are subjective and difficult to quantify due to the complexity of issues.

4.8   Identification of Indicators

Expert judgement was used along with extensive review of previous to select the indicators.

Datasets

A list of datasets used, their units, period of measurement, hypothesized relationships with vulnerability and data sources are summarized. A brief description of these datasets is given below.

4.8.1 Climatic Data

The study has used gridded monthly precipitation, maximum temperature and minimum temperature, time series data constructed by Climatic Research Unit (CRU TS 3.0) at a spatial resolution of 0.5 x 0.5 degree for the time period 1951 – 2009. The dataset was provided in the netCDF file format by BADC (http://badc.nerc.ac.uk). A program was written to extract netCDF format data into ENVITM format images as well as subset data for user given bounds through a GUI for further analysis. The images were geographically referenced automatically using the geographical extent read from the net CDF file. The gridded dataset of monthly terrestrial surface climate over land areas as constructed by New et al. (2000), and a detailed description of the datasets is given by Mitchell and Jones (2005) were used.

The gridded temperature data were used to calculate the rate of change over the years 1951-2009 by fitting the linear time trend, separately for kharif and rabi seasons. The monthly rainfall data were used to calculate Standardized Precipitation Index (SPI) (Mckee et al. 1993), an index of rainfall deviation for kharif and rabi seasons over the period 1951-2009. Since precipitation is not normally distributed, the long-term precipitation record was first fitted to an incomplete-gamma probability distribution, which was then transformed into a normal distribution to calculate SPI. The frequency of years when SPI was -1 or below was calculated for low rainfall, while the frequency of years when SPI was +1 and more was calculated for high rainfall. Besides, the severity of low rainfall was calculated by summing up all the SPI values whenever it was -1 or less for each grid. Similarly, the severity of high rainfall was calculated by summing up all the SPI values of +1 and more. The intensity of low and high rainfall was calculated by multiplying the corresponding frequency with modulus of severity. This way an index of rainfall variability was developed for kharif and rabi seasons. The district boundary layer was overlaid on rainfall intensity images and temperature trend images to compute district-wise average values.

4.8.2 Water-holding Capacity and Organic C Content of Soils

Available water-holding capacity (AWHC) of soil was estimated by taking the difference in water content between field capacity and permanent wilting point. The water-holding capacity of the soil mostly depends on soil porosity, which in turn, depends on soil texture, structure and bulk density. The organic C content of soil is an indicator of its fertility. The Digital Soil Map of the World and Derived Soil Properties (Version 3.5) produced by FAO (FAO, 1995) were used in this study. The soil map was produced at a finer resolution of $5' \times 5'$ cell size (9×9 km at equator) by using the World Inventory of Soil Emissions (WISE) database. The digital maps of soil moisture storage capacity (mm) for 1 m profile depth and soil organic carbon content (kg m^{-2}) were extracted for the study area. The district boundary layer was overlaid on it and the average values of soil moisture storage capacity and soil carbon content for each district were calculated using ArcGIS™.

4.8.3 Irrigated Area

The study utilized the FAO Global Map of Irrigation Areas (GMIA) version 4.0.1 having a cell size of $5' \times 5'$ (Siebert et al. 2007) for calculating the irrigated area. Each cell of the map depicts the area quipped for irrigation as the percentage of cell area around the year 2000, but for India it refers to the area actually irrigated. The global map of irrigated areas was developed by combining sub-national irrigation statistics with geospatial information on the position and extent of irrigation schemes to compute the fraction of arc-minute cells that were equipped for irrigation, is called irrigation density. A more detailed description of the dataset, development and validation is given in Siebert et al. (2005). The global ASCII data were unzipped, imported into

ENVI™, geo-coded and sub-setted for the study area. The district layer was overlaid on it and the average value of percent irrigated area was calculated.

4.8.4 Agricultural Statistics

The district-wise statistics on net sown area and geographical area were compiled from the Census of India (2001). District-wise productivity of food grains was obtained from the Agricultural Statistics published by the Directorate of Economics and Statistics, Department of Agriculture and Cooperation, Ministry of Agriculture, Government of India for the period 2001 to 2006. Livestock density was compiled from the 17th Livestock Census (2003) published by the Department of Animal Husbandry and Dairying, Government of India. The average size of landholding and cropping intensity statistics were compiled from the reports published by the Departments of Agriculture of respective states. The annual N, P and K fertilizer consumption statistics were obtained from the Fertilizer Statistics published by the Fertilizer Association of India.

4.8.5 Socio-economic Statistics

The district-wise statistics of human population density, number of villages electrified and the number of villages with paved roads were compiled from the Census of India (2001). The human development index (HDI) was obtained from the Human Development reports of respective states, as produced by the UNDP. Two types of functional relationships are possible between the indicators and vulnerability. The vulnerability is directly proportional to the value of some indicators and inversely proportional to the value of some other indicators. For example, for indicators such as change in maximum and minimum temperature, the higher is the value of these indicators; more will be the vulnerability of the region to climate change as variation in climatic variables increases the vulnerability of agriculture. Thus, the indicators have positive functional relationship with vulnerability. On the other hand, for the HDI, a higher value implies more education, better health, and more income resulting in more awareness to cope with the climate change. As a result, the vulnerability will be lower and thus the HDI has a negative functional relationship with vulnerability.

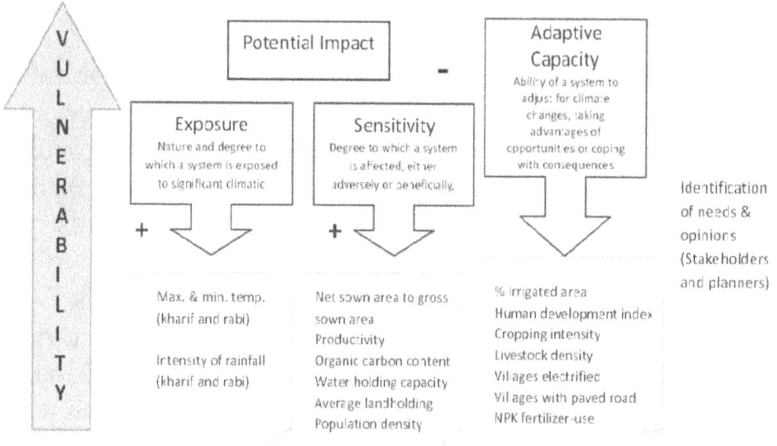

Fig. 4.3 Indicators of vulnerability used in the study

4.9   Ranking of Indicators

Each indicator of exposure, sensitivity and adaptive capacity was classified into 5 classes to assign the ranks. In this study, the five-point ordered scale was used to rank each factor from very low to extreme value. The value 0 was left in scaling to define mask value. Thus, each factor was having an equivalent measurement basis or scale before any weight was applied. The ranks were assigned to the indicators according to their functional relationship with vulnerability, i.e., if the indicator was directly related to vulnerability; higher ranks were given for higher values. However, in case the indicator was inversely related, lower ranks were given for higher values.

4.10   Calculation of Vulnerability Index

Once the weight of each indicator was determined, exposure, sensitivity and adaptive capacity maps were prepared by taking weighted some of the rank of all relevant indicators. These three parameters and maps along with their functional relation with vulnerability resulted in the final calculation of vulnerability map varying from 1 to 4 (low, moderate, high and extreme). The composite vulnerability rating maps were produced in GIS.

References

Gallopin GC (2003) A systemic synthesis of the relations between vulnerability,hazard, exposure and impact, aimed at policy identification. In Economic Commission for Latin American and the Caribbean (ECLAC). Handbook for estimating the socio-economic and environmental effects of disasters. Mexico, D.F.: ECLAC.

IPCC (Intergovernmental Panel on Climate Change) (2001) The Regional Impacts of Climate Change: An Assessment of Vulnerability: Special Report of IPCC Working Group II (Watson, R.T., M.C. Zinyowera, and R.H.Moss (eds.). Intergovernmental Panel on Climate Change, Cambridge, UK: Cambridge University Press. 517.

IPCC (Intergovernmental Panel on Climate Change) (2001) Climate Change 2001: Impacts, Adaptation, and Vulnerability, McCarthy JJ, et al. eds. Contribution of Working Group II to the Third Assessment Report of the Intergovernmental Panel on Climate Change. Cambridge, UK: Cambridge University Press.

IPCC (Intergovernmental Panel on Climate Change) (2007) Climate Change 2007: Impacts, Adaptation and Vulnerability. Contribution of Working Group II to the Fourth Assessment Report of the Intergovernmental Panel on Climate Change, Parry ML, Canziani OF, Palutikof JP, van der Linden PJ and Hanson CE, eds. Cambridge, UK: Cambridge University Press. Malczewski J (1999) GIS and Multi-criteria Decision Analysis. Toronto, Canada.John Wiley & Sons.

McCarthy JJ, Canziani OF, Leary NA, Dokken DJ and White KS, eds. (2001). Climate change 2001: Impacts, Adaptation and Vulnerability. Cambridge, UK: Cambridge University Press.

Mckee TB, Doesken NJ and Kleist J (1993) The relationship of drought frequency and duration to time scales. In: Proc. 8th conference on Applied Climatology, Anaheim, California. p. 179-184.

Mitchell TD and Jones PD (2005) An improved method of constructing a database of monthly climate observations and associated high-resolution grids. Int. J. Climatol. 25:693-712.

Mckee TB, Doesken NJ and Kleist J (1993) The relationship of drought frequency and duration to time scales. In: Proc. 8th conference on Applied Climatology, Anaheim, California. p. 179-184.

New M, Hulme M and Jones PD (2000) Representing twentieth century space-time climate variability. Part 2: development of 1901-96 monthly grids of terrestrial surface climate. J. Climate 13:2217-2238.

SEI (Stockholm Environment Institute) (2004) Choosing methods in assessments of vulnerable food systems. Risk and Vulnerability Programme, SEI Briefing Note, Stockholm: SEI.

Watson RT, Zinyowera MC and Moss RH (eds) (1996) Climate change 1995: Impacts, Adaptations and Mitigation of Climate Change: Scientific-Technical Analyses, contribution of Working Group II to the second assessment report of the Intergovernmental Panel on Climate Change. Cambridge, UK: Cambridge University Press. p. 38.

5

Need for Building Capacities of Institutions and Individuals to Deal with Climate Change

G. V. Ramanjaneyulu

Centre for Sustainable Agriculture (CSA), Hyderabad.

5.1   Preamble

Agriculture and Climate Change are mutually impacted. Often the impact of climate change is underestimated, and the contribution of agriculture to Climate Change is ignored. As a result much of the discussion, debates on climate change and agriculture are around particular technologies which can help farming to adapt to climate change. In reality, if farmers have to adapt to the changing climate, we need to understand this in a broader context of ecological, economical and socio political crisis which Indian farmers are already undergoing.

Climate change is already a reality for a majority of Indian farmers even as plans are being evolved at the government level mostly to create adaptive capabilities; meanwhile, Indian farmers are being forced to adapt to several CC-related changes by themselves because they have no other choice[1]. For no fault of theirs, Indian farmers, like the most marginalized everywhere, are paying a high price for anthropogenic climate change. The worst-hit, as usual, are small and marginal holders in marginalized locations with social disadvantages to begin with. Such farmers have meagre resources to buffer them from the new risks that climate change poses.

---

[1] Sustaining Agriculture in the era of Climate Change in India - Civil Society position paper http://agricoop.nic.in/Climatechange/ccr/files/Sustainable%20Agriculture-OXFAM-Position_paper-CSA.pdf

The relationship between climate change and agriculture is three folds. First, climate change has a direct bearing on the biology of plant and animal growth. Second, the changes in the farm ecology – such as, for example soil conditions, soil moisture, pests and diseases etc. Third the ability of the existing social and economic institutions, particularly in rural areas, to deal with the challenges posed by global warming. In the larger context of food security and climate change, it is also important to consider other sectors like animal husbandry and livestock, which are closely linked with agriculture[2].

5.2   Understanding India in the Larger Context of Climate Change

Though India occupies only 2.4% of Earth's surface area, is home of 17.5% of the world's population. Between 2001 and 2011 India has added 181 million people to the world population. The most recent national census projects that by 2025 India will surpass China to become the world's most populous nation. By 2050, India's population is predicted to reach 1.6 billion. About 60 per cent of the rural households were farmer households engaged in farming activities like cultivation, plantation, animal husbandry, fishery, bee-keeping and other agricultural activities (NSSO, 2003)[3]. As per the data from the Census Division, Ministry of Agriculture, Government of India, operational holdings below 4.0 hectares (ha) constitute 93.6% of the operational holdings covering 62.96% of the operational area, or about 100.65 million ha in absolute terms (Agricultural Census, 2005)[4].   The looming economic crisis in terms of increased cost of cultivations,

---

[2]Women's Perspective on Climate Change Adaptation https://kspcb.files.wordpress.com /2015/04/womens-perspective-climate-change_pacs-letter-head.pdf

[3]Climate Change and Indian Agriculture: Implications and Way Forward https://in.boell.org/sites/default/files/climate_change_and_indian_agriculture_imlpcations_way_forward_book.pdf

[4]Sustaining Agriculture in the era of Climate Change in India-Civil Society position paper   http://indiaforsafefood.in/PDF/civil_society_position_paper_with_coverpage.pdf

decreasing incomes and ecological crisis due to high use of chemicals, degrading soils and receding ground water levels would further be aggravated by the uncertainties posed by the climate change[5].

India is faced with the challenge of sustaining its rapid economic growth while dealing with the global threat of climate change. This threat emanates from the accumulated green house gas emissions in the atmosphere, anthropogenically generated through long term and intensive industrial growth and high consumption lifestyles in developed countries.

5.3   Impact of Climate Change on Agriculture

India being predominantly under rainfed agriculture may be more prone to the impacts of climate change. Climate change is also poised to have a sharply differentiated effect as between agro-ecological regions, farming systems, and social classes and groups. Rainfall extremities are being witnessed frequently. For instance, it is reported that about two-thirds of the sown area in the country is drought-prone and around 40 million hectares is flood-prone. The poorest people are likely to be hardest hit by the impacts of climate variability and change because they rely heavily on climate-sensitive sectors such as rainfed agriculture and fisheries. They also tend to be located geographically in more exposed or marginal areas, such as flood plains or nutrient-poor soils. The poor also are less able to respond due to limited human, institutional and financial capacity and have very limited ability to cope with climate impacts and to adapt to a changing hazard burden.

Climate change is manifesting itself in many ways across the country. Among the indicators, while long term rainfall data analysis shows no clear trend of change, regional variations as well as increased rainfall during summer and reduced number of rainy days can be noticed. In the case of temperature, there is a 0.6 °C rise in the last 100 years and it is projected to rise by 3.5-5 °C by 2100. The carbon dioxide concentration is increasing by

---

[5]Global Trends And Future Challenges For The Work Of The Organization
http://www.fao.org/docrep/meeting/025/md883E.pdf

1.9 ppm each year and is expected to reach 550 ppm by 2050 and 700 ppm by 2100. Extreme events like frequency of heat and cold waves, droughts and floods have been observed in the last decade. The sea level has risen by 2.5 mm every year since 1950 while the Himalayan glaciers are retreating. These are all symptomatic of climate change.

Available research indicates that climate change-induced rise in temperature is going to affect rainfall patterns – farming in India depends on monsoons and there is a close link between climate and water resources.

The organic carbon levels and moisture in the soil will go down while the incidence of runoff erosion will increase. The quality of the crop may also undergo change with lower levels of nitrogen and protein and an increased level of amylase content. In paddy, zinc and iron content will go down which will impact reproductive health of animals. Insect lifecycles will increase which in turn will raise the incidence of pest attacks and virulence. Other likely impacts are change in farm ecology viz. bird-insect relations, and an increase in the sea levels which will cause salinity ingression and submergence.

It is projected that due to climate change, kharif rainfall is going to increase and this might be positive for kharif crops. Further, for *kharif* crops, a one-degree rise in temperature may not have big implications for productivity. However, temperature rise in rabi season will impact production of wheat, a critical food-grain crop.

The surface air temperatures will increase by 2 to 4 °C by 2070-2100. As mentioned earlier, the *rabi* crop will be impacted seriously and every 1 °C increase in temperature reduces wheat production by 4-5 million tons, as per a study by IARI. This loss can be reduced to 1-2 million tons only if farmers change to timely planting. Increased climatic extremes like droughts and floods are likely to increase production variability. Productivity of most cereals would decrease due to increase in temperature and decrease in water availability, especially in Indo-Gangetic plains. The loss in crop production is projected at 10-40% by 2100, depending upon the modeling technique applied.

The impacts of climate change are already visible. A network of 15 centres of ICAR working on studying climate change has reported that apple production is declining in Himachal Pradesh due to inadequate chilling. This is also causing a shift in the growing zone to higher elevations. Similarly, in the case of marine fisheries, it has been observed that Sardines are shifting from the Arabian Sea to the Bay of Bengal, which is not their normal habitat. In fact, fisheries are the most vulnerable sector to climate change[6]. Crops have the ability to adapt to extreme climate variability even up to, say, 4 °C while fishes and animals do not. It has also been recorded that the pest ecology of certain crops is changing due to climate change.

Extreme weather conditions resulting in disasters will obviously have their own socio-economic impacts, especially on the poor. Further, changes in crop productivity will have implications on farmers' incomes.

5.3.1 Pest and Disease Shift

Insect populations like all animal populations are governed by their innate capacity to increase as influenced by various abiotic and biotic factors. The changes caused by the natural evolutionary forces are accelerated with the human interventions. After the changes like depletion of natural resources, environmental pollution, extinction of certain species of plants and animals, the Climate Change particularly caused by inadvertent anthropogenic disturbances has become more evident. Projected carbon dioxide levels in the twenty first century are two to four times higher than the pre industrial era. Weather and Climate have an impact on the pest population. Climate change leads to shifts in the pest incidence, migration and viability thresholds. Today, farming and farmers lives are already effected by the pests and pest management practices they adopt. Hence, understanding the intricacies of climate change on pest management in agriculture is crucial.

---

[6]Farming under the changing skies http://www.goimonitor.com/story/ farming- under-changing-skies

Weather has both direct and indirect effects on the insect populations. Reproductive biology of an insect may be affected both positively and negatively which ultimately affect its populations. Climate change would shift pest populations as it would shift important determinants of pest incidence, namely temperature, precipitation distribution, and wind pattern.

Temperature: All life survives with a certain narrow range of temperature. Deviations from this in optimum range on either side are tolerated to some extent, depending on the physiological adaptations of the concerned species or populations. Temperatures above or below these limits can prove lethal. Exposure to lethal high or low temperatures may result in instant killing or failure to grow and reproduce normally. Harmful effects of exposure to sub-lethal temperatures may be manifested at later critical stage like moulting or pupation. The rise in temperature might also have a negative effect on delicate natural enemies such as hymenopteran parasitoides and small predators. This may affect natural enemy-pest relationship. For e.g., Brown Plant Hopper is 17 times more tolerant to 40 ºC than its predator *Cyrtorrhynus lividipennis* but wolf spider *Paradosa pseudoannulata* is tolerant to 40 ºC.

Moisture: Most terrestrial insects live in an environment, which is dry. The only source of water for insects is the water obtained with food material from their host plants. These insects have, therefore developed a variety of mechanisms to conserve water. In spite these mechanisms, exceptionally dry air may prove lethal to most insects. Likewise, excessive moisture may also adversely affect many insects by encouraging disease outbreaks, affecting normal development and by lowering their capacity to withstand lower temperatures. The reproductive capacity of the insects is also affected by moisture but there are great differences in the capacity of different insects to tolerate conditions ranging from extreme dryness to near saturated environments. For example, incidence of Rice Hispa in Telangana region has increased in the last two years due to prevailing dry situations.

There is a shift observed from the leaf/fruit eating caterpillars to sucking pests in the recent years. While monoculture of crops/varieties and chemical pest management practices understood to have resulted in such pest shifts, climate change also have contributed for such shift. For example in cotton there is a shift towards sucking pests (mealy bugs, jassids) particularly after the introduction of Bt cotton. Similarly, Aphid incidence in Groundnut, Thrips and yellow mites in chillies are observed. Most of the sucking pests are also vectors of viral diseases. With increasing incidence of sucking pests viral diseases are also increasing e.g., Budnecrosis in Groundnut, Tobacco Streak Virus incidence in Cotton, and similar viral problems in most of the fruit crops, vegetables.

5.4   Impacts of Agriculture on Climate Change

Agriculture contributes around 10-12% of total global greenhouse gas (GHG) emissions but is the main source of non-carbon dioxide (CO_2) GHGs emitting nearly 60% of nitrous oxide (N_2O) and nearly 50 % of methane (CH_4) (Smith et al., 2007a)

Capital depletion and massive additions of external inputs in a 'linear' model (as opposed to cyclical systems in ecological farming models) are characteristic features of intensive, industrialized models of agriculture. In this model, a farm is treated like a factory with inputs and outputs calculated in a monocropped situation with grain yield given the highest importance compared to any other parameter and often, the externalized costs are ignored.

GHGs and their Global Warming Potential

Measure of the ability of a gas in the atmosphere to trap heat radiated from the earth's surface compared to a reference gas, which is usually assumed to be carbon dioxide

- carbon dioxide (CO_2) = 1;
- methane (CH_4) = 21;
- nitrous oxide (N_2O) = 310;
- sulphur hexafluoride (SF_6) = 23,900;
- tetrafluoromethane (CF_4) = 6500;
- hydrofluorocarbons (HFCs): HFC-134a = 1300;
- chlorofluorocarbons (CFCs): CFC-114 = 9300
- hydrochlorofluorocarbons (HCFCs): HCFC-22 = 1700 IPCC (2007)

Amongst various GHGs that contribute to global warming, carbondioxide is released through agriculture by way of burning of fossil fuel; methane is emitted through agricultural practices like inundated paddy fields, for example; nitrous oxide through fertilizers, combustion of fossil fuels etc. Nitrous oxide has a global warming potential 296 times greater than CO_2. In India, it is estimated that 28% of the GHG emissions are from agriculture; about 78% of methane and nitrous oxide emissions are also estimated to be from agriculture.

5.4.1 Chemical Fertilizers and Climate Change

Nitrogen fertiliser manufacture and application to the soil contribute significantly to greenhouse gases (GHG) emissions and thus, climate change. India consumes ~14 Mt of synthetic N every year, of which about 80 per cent is produced within the country, making it the second largest consumer and producer of synthetic N fertiliser in the world, after China. As per an estimate the GHG emissions from fertiliser manufacture and use in India reached nearly 100 million tonnes of CO_2 eq in 2006/07, which represents about 6 percent of total Indian greenhouse gas emissions (Roy et.al 2010).

- An estimate of emissions from the manufacture of synthetic N fertilizers

- Manufacture of synthetic nitrogen fertilizer is a very energy intensive process, and currently requires large amounts of fossil fuel energy.

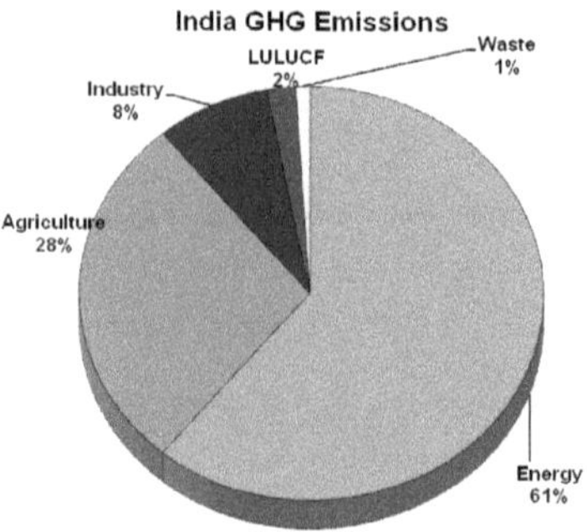

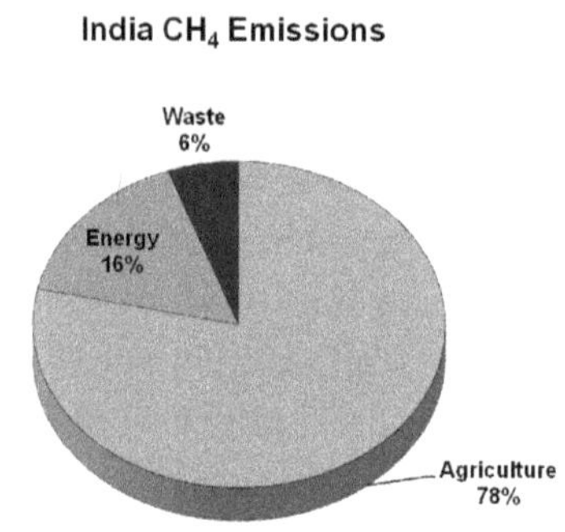

- Natural gas is the main fuel and feedstock, which accounts for 62 per cent of the energy used in synthetic N fertilizer production.

- Less efficient and more polluting fuels such as naphtha and fuel oil also represent a high share, 15 and 9 per cent respectively, of the energy used in fertilizer manufacture (values as of 2006/07, FAI 2007).

- Of the various forms in which synthetic N fertilizers are available, urea accounts for a chunk of the total N fertilizer produced and consumed (81 per cent in 2006).

- The synthesis of urea is based on the combination of ammonia and CO_2 and its emissions are dominated by CO_2

- While other synthetic N fertilizers comprise a smaller percentage of the fertilizer market, they make notable emissions to the atmosphere both during production and consumption. We calculated emissions from the manufacture of synthetic N fertilizer following the Intergovernmental Panel on Climate Change (IPCC) methodology.

- Total **greenhouse gas emissions** (GHG) from the manufacturing and transport of fertiliser are estimated at **6.7 kg CO_2 equivalent** (CO_2, nitrous oxide and methane) **per kg N**

- Globally, an average **50%** of the nitrogen used in farming is lost to the environment. Significant amounts escape into the air, or seep into the soil and underground water, which in turn result in a host of environmental and human health problems, from climate change and dead zones in the oceans to cancer and reproductive risks (Galloway et al., 2008)

- 1.25 kg of N_2O emitted per 100 kg of Nitrogen applied

- as nitrate polluting wells, rivers, and oceans

- Volatilization loss 25-33 %

- Leaching loss 20-30 %

- Being high energy intensive, the fertilizer prices increase as the feed stock prices rise. The increased costs are subsidised by the Central Government and the subsidy reached Rs. 90,000 crore in Indian Budget during 2011-12 as per the revised estimates.

- After Nutrient Based Subsidy was introduced in 2008, fertilizer prices were decontrolled except for urea and prices have increased by five folds.

- With the Phosphotic reserves in the world are depleting and could be economically be exploited only for another 25 years.

5.4.2 Burning Crop Residues

Another major contributor of GHGs is the burning of crop residues. In Punjab, wheat crop residue from 5,500 square kilometers and paddy crop residues from 12,685 square kilometers are burnt each year. Every 4 tons of rice or wheat grain produces about 6 tons of straw. Emission Factors for wheat residue burning are estimated as: CO- 34.66 g/Kg, NOx – 2.63 g/Kg, CH_4 – 0.41g/Km, PM10 – 3.99 g/Kg, PM2.5 – 3.76 g/Kg.

Burning of crop residues also impacts the soil (fertility). Heat from burning straw penetrates into the soil up to 1 cm, elevating the temperature as high as 33.8–42.2°C. Bacterial and fungal populations are decreased immediately and substantially in the top 2.5 cm of the soil upon burning. Repeated burning in the field permanently diminishes the bacterial population by more than 50%. The economic loss due to the burning of crop residues is colossal. Each year 19.6 million tonnes of straw of rice and wheat, worth crores of rupees are burnt. Used as recycled biomass, this potentially translates into 38.5 lakh tonnes of organic carbon, 59,000 tonnes of nitrogen, 2,000 tonnes of phosphorous and 34,000 tonnes of potassium every year.

5.4.3 Inundated Paddies

Another potent GHG is methane which is emitted in copious amounts through inundated paddy cultivation. Rice paddies emit CH_4 when they are flooded due to the anaerobic decomposition of organic matter in the soil producing the gas, which then escapes to the atmosphere mainly through diffusive transport through the rice plants (Nouchi et al., 1990), or is oxidized before reaching the surface. The level of CH_4 emission from any given rice paddy is related to factors that control the activity of the methane producing (methanogens) and methane-oxidizing bacteria (methanotrophs) such as temperature, pH, soil redox potential and substrate availability, and also soil type, rice variety, water management and fertilization with organic carbon and N (see reviews by Le Mer and Roger, 2001, and Conrad, 2002).

In India, of a total area of 99.5 Mha under cereal cultivation, 42.3 Mha (or 42.5%) is under rice cultivation. It is grown under flooded conditions and the seedbed preparation involves puddling or plowing when the soil is wet to destroy aggregates and reduce the infiltration rate of water. Such anaerobic conditions lead to emission of methane and possibly nitrous oxide through inefficient fertilizer use.

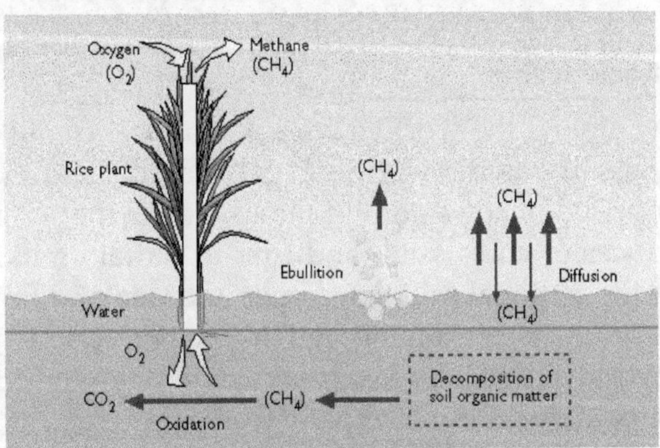

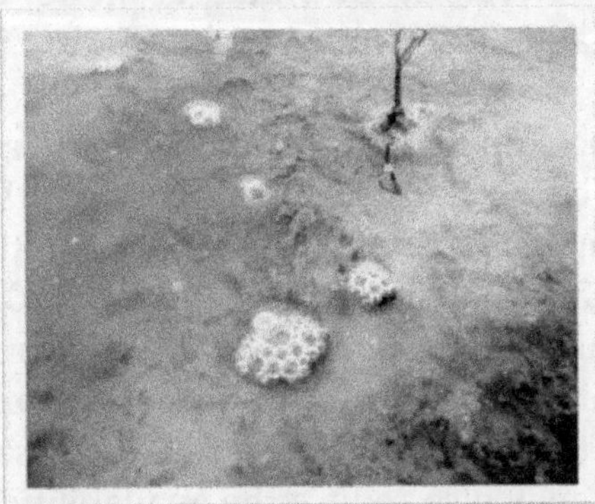

Emission of methane from rice paddies in India is differentially estimated. The average methane flux from rice paddies ranges from 9 to 46 g/m² over a 120- to 150-day growing season

CH$_4$ emissions (Tg CO$_2$ –eq yr^{-1})	References
55.2-138	Parashar *et al*, 1994
135	Yan *et al*, 2003
94.07 ± 27.37	Gupta *et al.*, 2009

5.4.4 Large Dams and GHGs

Another indirect contribution of agriculture to GHG emissions comes in the form of large dams. Large dams contribute 18.7% of emissions in India as per an estimate. Total methane emissions from India's large dams could be 33.5 million tonnes (MT) per annum, including emissions from reservoirs (1.1 MT), spillways (13.2 MT) and turbines of hydropower dams (19.2 MT).

5.4.5 Livestock Sector

India is now among the world's largest producer of milk, poultry, meat and eggs. It has the world's biggest dairy herd, 300 million strong, comprised of cows and buffaloes, and is the second largest global producer of cows' milk and first in buffalo milk. It is also the world's top national milk consumer and demand for milk and other dairy products is growing by 7 to 8% per year. This country is also the world's fourth largest producer of eggs and fifth largest producer of poultry meat, principally from chicken.

However, the livestock in India is more distributed and household based and mostly integrated with crop production. The crop residues are used as fodder and the animal waste is used as the manure for the crop fields. The impacts of livestock on climate change needs to be understood in this context. Livestock is also impacted by climate change. Possible temperature increases in India of between 2.3 to 4.8 degrees Celsius by 2050 will add to heat stress in animals used to produce milk and affect reproduction and the amounts of milk each animal provides. Crossbred cows may be most vulnerable to higher temperatures. Increased temperatures and sea level rise may also reduce the availability of land to grow feed, and result in lower crop yields and an increase in the severity and spread of animal diseases.

In 2010, India was the world's fastest growing poultry market, outpacing Brazil, China, the US and the European Union and Thailand. The costs of producing chicken for meat in the country is world's second lowest and production of eggs in India is cheaper than in any other country, according to the Poultry Federation of India. India is the top global exporter of buffalo meat and it is also exports increase quantities of maize and soy, both important ingredients in commercial feed. In addition, India's leading poultry producers are expanding their sales to countries in Asia and Middle East.

Greenhouse gases (GHGs) are generated at virtually every point along the livestock production chain. The UN Food and Agriculture Organization (FAO) attributed to the global livestock industry 9% of the anthropogenic (man-made) CO_2 emissions; 37% of anthropogenic methane emissions, and 65% of anthropogenic nitrous oxide emissions. Methane and nitrous oxide, though lower in concentration in the atmosphere than CO_2, are far more potent heat-trapping gases. In India, emissions from the energy used by agriculture and fisheries industries totaled 34 million tons of CO_2 or 3% of the GHGs produced by the energy sector.

This does not include emissions from electricity taken from the national grid to activities such as cool large poultry or egg operations or dairies, or to slaughter and process animals and their products. Soil cultivation related to animal agriculture globally emits about 28 million tons of CO_2 every year. More than half of this energy used in producing milk and eggs can be attributed to feed production. There are other indirect CO_2 emissions, specifically from the manufacture of chemical and nitrogen based fertilizers. About 41 million tons of CO_2 are emitted globally each year from the production of nitrogen fertilizers applied to feed crops.

Carbon dioxide is also released when forests and other vegetation are destroyed to make way for feed crops or pasture. Considerable uncertainty exists in calculating overall GHGs from such changes in land use, though the FAO estimates that 2.4

billion tons of CO_2 are emitted every year due to deforestation to create pasture land for livestock or land for cultivation of feed crops. On the top of this, 100 million tons of CO_2 is released every year from livestock-induced desertification of land.

Other Farm Operations

GHG Emission from the Use of Farm Machinery

The other major source of energy emissions is also the issue of more energy use in intensive farming models in the form of fossil fuels for machinery like tractors, harvesters and so on, pumps for irrigation etc.

Operation type	Emission level (kg CO_2 –eq ha^{-1})
Tillage	4.40-73.60
Drilling or Seeding	8.10-14.30
Application of Agrochemicals	1.80-37.00
Combine harvesting	2210-42.10

Calculated from data in Lal (2004)

5.5   Adapting to Climate Change

While global warming and the resultant climate change have significant impacts on agriculture, agriculture based livelihoods and food security, many practices in agriculture also contribute to climate change and there are several other practices under the broad category of 'sustainable agriculture' that also have the potential to adapt to or mitigate such GHG emissions otherwise occurring from farming.

Approaches to Climate Change Adaptation: The two main types of adaptation are autonomous and planned adaptation. Autonomous adaptation is the reaction of, for example, a farmer to changing rainfall patterns, in that say changes crops or uses different harvest and planning/sowing dates, by trial and error.

Planned adaptation measures are conscious policy options or response strategies, often multisectoral in nature, aimed at altering the adaptive capacity of the agricultural system or facilitating specific adaptations. For example, deliberate crops/varieties selection, promoting/discouraging certain practices by incentivizing/regulating etc., when ever the adaptation measures are considered holistically including trade-offs among biophysical and socio-political factors.

While global warming and the resultant climate change have significant impacts on agriculture, agriculture based livelihoods and food security, many practices in agriculture also contribute to climate change and there are several other practices under the broad category of 'sustainable agriculture' that also have the potential to adapt to or mitigate such GHG emissions otherwise occurring from farming.

Biodiversity in all its components (e.g., genes, species, ecosystems) increases resilence to changing environmental conditions and stresses. Genetically-diverse populations and species rich ecosystems have greater potential to adapt to climate change. Use of indigenous and locally-adapted plants and animals, hence, selection and multiplication of crop varieties animal species locally adapted and resistant to adverse conditions is essential.

Work on adapted crops and animals cannot be separated from their management options within agro-ecosystems. For example, Rice one the staple food crops of India had several varieties with different abilities to tolerate high temperature, salinity, drought and floods. Rice varieties with salinity tolerance have been used to expedite the recovery of production in areas damaged by the 2004 tsunami (FAO, 2007). Similarly, practices like System of Rice Intensification can reduce the water usage and thereby methane emissions from the paddy fields. It was observed that the methane emissions are 4 times lesser and Nitrous oxide emissions are 5 times lesser from SRI fields compared to conventional paddy fields (Karki 2010).

Climate change adaptation for agricultural cropping systems requires a higher resilience against both excess of water (due to high intensity rainfall) and lack of water (due to extended drought periods). A key element to both problems is soil organic matter, which improves and stabilizes the soil structure so that the soils can absorb higher amounts of water without causing surface runoff, which could result in soil erosion and further downstream, in flooding. Soil organic matter also improves the water absorption capacity of the soil for during extended drought. While intensive tillage reduces soil organic matter through aerobic mineralization, low tillage and the maintenance of a permanent soil cover (through crops, crop residues or cover crops and the introduction of diversified crop rotations) increases soil organic matter. A no- or low-tilled soil conserves the structure of soil for fauna and related macropores (earthworms, terites and root channels) to serve as drainage channels for excess water. Surface mulch cover protects soil from excess temparatures and evaporation losses and can reduce crop water requirements by 30 per cent. Thus organic/ecological farming can increase soil organic carbon, reduce mineral fertilizers use and reduce on-farm energy costs.

A broad range of agricultural water management practices and technologies are available to spread and buffer production risks. Enhancing residual soil moisture through land conservation techniques assists significantly at the margin of dry periods while buffer strips, mulching and zero tillage help to mitigate soil erosion risk in areas where rainfall intensities increase. The inter-annual storage of excess rainfall and the use of resource efficient irrigation remain the only guaranteed means of maintaining cropping intensities.

Sustainable   Agriculture   (ecological   farming/organic farming/LEISA/Non Pesticidal Management/SRI etc) approaches are now acknowledged for the wide set of ecological and economic benefits that accrue to the practitioners as well as consumers of agricultural products. These approaches which are based on low external inputs are also low energy intensive and less polluting hence mitigate and help in adapting to the climate change.

However, the promotion of sustainable agriculture on a large scale is often confronted about its potential as well as its practical limitations. In the last five years two large scale initiatives, NPM scaling up (Community Managed Sustainable Agriculture-CMSA) in Andhra Pradesh and SRI promotion in states of Tripura, Orissa and Tamil Nadu have brought in new learnings and broken the earlier apprehensions on scaling up such practices and their relevance on a large scale.

These successful experiences had three elements in common. First, all have made use of locally adapted resource conserving technologies. Second, in all there has been coordinated action by groups or communities at local level. Third, there have been supportive external (or non-local) government and/or non-governmental institutions working in partnership with farmers. Almost every one of the successes has been achieved despite existing policy environments which still strongly favor 'modern and established' approaches (technology and support systems) to agricultural development.

Conclusions

Now the challenge is how these can be scaled up onto a large scale across the nation given the wide diversity of situations. This needs a newer approach in terms of capacity building, horizontal learning, newer institutional systems and newer forms of financial support to be put in place. The programmatic support to agriculture today favour only high external input based agriculture. As a result, none of the mainstream programs provide any support for promotion of these models. This needs the recasting of program guidelines or initiating newer program to provide support to more sustainable models in agriculture which can be easily accessible to small and marginal farmers. The mission on sustainable agriculture can initiate a programmatic support to scale up sustainable agriculture with the objectives to

- Reduce the risks with uncertain weather conditions and degraded and limited natural resources in these regions, by adopting suitable cropping patterns and production practices,

- Diversify the assets and income sources to sustain the livelihoods by integrating livestock and horticulture into agriculture and promoting on-farm and off-farm employment opportunities,

- Conserve and efficiently use the available natural resources like soil and water, and promote biomass generation,

- Organize farmers into institutions which can help them to have better planning, greater control over their production, help to access resources and support, improve food security and move up in the value chain,

- Build livelihood security systems to withstand the natural disasters like drought, floods and other climate uncertainties.

References

Conrad, R. (2002). Control of microbial methane production in wetland rice fields. Nutrient Cycling in Agro ecosystems 64: 59-69.

Galloway, J., Raghuram, N. and Abrol, Y.P. (2008). A perspective on reactive nitrogen in a global, Asian and Indian context. Current Science 94 (11): 1375-1381.

Gupta, P.K., Gupta, V., Sharma, C., Das, S.N., Purkait, N., Adhya, T.K., Pathak, H., Ramesh, R., Baruah, K.K., Venkatratnam, L., Gulab Singh and Iyer,C.S.P. (2009). Development of methane emission factors for Indian paddy fi elds and estimation of national methane budget. Chemosphere 74, 590–598.

IPCC, AR. (2007). Intergovernmental panel on climate change."Climate change 2007: Synthesis report.

Jain, P.C. and Bhargava, M.C. (2007). Entomology: Novel Approaches. New India Publishing, New Delhi.

Karki, S. (2011). System of Rice Intensification: an analysis of adoption and potential environmental benefits.

Le Mer, J. and Roger, P. (2001). Production, oxidation, emission and consumption of methane by soils: A review. European Journal of Soil Biology 37, 25-50.

Nguyen, N.V. (2007). Global climate changes and rice food security. Ed, Volume by FAO, Rome, Italy.

Nouchi, I., Mariko, S. and Aki, K. (1990). Mechanism of methane transport from the rhizosphere to the atmosphere through the rice plants. Plant Physiology 94:59–66.

Parashar, D.C., Mitra, A.P., Sinha, S.K., Gupta, P.K., Rai, J., Sharma, R.C., Singh, N., Kaul, S., Lal, G., Chaudhary, A., Ray, H.S., Das, S.N., Parida, K.M., Rao, S.B., Kanungo, S.P. (1994). Methane budget from Indian paddy fi elds. In: K. Minami, A. Mosier and R. Sass (Eds). CH4 and N2O: Global Emissions and Controls from Rice Fields and Other

Agricultural and Industrial Sources. NIAES. Yokendo Publishers, Tokyo, pp. 27-39.

Ramanjaneyulu, G. V., and Dr. Raghunath, T. A. V. S. (2009). Climate change and Pest Management in Agriculture. http://ramoo.in/?p=143

Roy, B.C., Chattopadhyay, G.N. and Tirado, R. (2009). Subsidising Food Crisis. Synthetic Fertilizers lead to poor soil and less food. Greenpeace India, 6. Bangalore

Smith, P., Martino, D., Cai, Z., Gwary, D., Janzen, H.H., Kumar, P., McCarl, B., Ogle, S., O'Mara, F., Rice, C., Scholes, R.J., Sirotenko, O., Howden, M., McAllister,T., Pan, G., Romanenkov, V., Rose, S., Schneider, U. and Towprayoon, S.(2007). Agriculture. Chapter 8 of Climate change 2007: Mitigation. Contribution of Working Group III to the Fourth Assessment Report of the Intergovernmental Panel on Climate Change [B. Metz, O.R. Davidson, P.R. Bosch, R. Dave and L.A. Meyer (Eds)], Cambridge University Press, Cambridge, United Kingdom and New York, NY, USA.

Tirado, R., Gopikrishna, S.R., Krishnan, R. and Smith, P. (2010). Greenhouse gas emissions and mitigation potential from fertilizer manufacture and application in India. International Journal of Agricultural Sustainability, 8(3): pp.176-185.

Yan, X.Y., Ohara, T. and Kimoto, H. (2003). Development of region-specific emission factors and estimation of methane emission from rice fields in the East, Southeast and South Asian countries. Global Change Biology 9, 237-254.

6

Impact of Climate Change on Soil Properties and their Mitigation Strategies

Pragati Pramanik

*Centre for Environment Science and Climate Resilient Agriculture
Indian Agricultural Research Institute (IARI), New Delhi.*

6.1 Introduction

The whole climatic system is changing due to changing levels of C and N based greenhouse gases in the atmosphere. C and N are both important components of soil organic matter and soil being a part of C and N cycles will be affected by climate change. Changes in average temperatures and in precipitation patterns will affect soil organic matter which, in turn will alter important soil physical and chemical properties like aggregate formation and stability, water holding capacity, cation exchange capacity, microbial diversity and soil nutrient content. As per recent IPCC report, the average global temperature will increase between 1.1 and 6.4 °C by 2090–2099 as compared to 1980–1999 temperatures (IPCC, 2007). Climate change will alter global precipitation patterns, both the amount and the distribution of precipitation in many locations [IPCC, 2007]. This change in climate will affect the environment, including the soil [Brevik, 2012] through its effects on soil properties and processes (Brevik, 2013). Therefore, this paper will focus on how climate change may alter soil properties and processes.

6.2 Climate Change and its Impact

Temperature: The average earth surface temperature has increased by 0.6 °C since the late nineteenth century. The 1990s was probably the warmest decade and 1998 the warmest year since 1860 IPCC, 2001). As per Fourth Assessment Report of IPCC (2007), world temperatures could increase by 1.1 to 6.4 °C during the 21st century.

Rainfall and humidity: Over the sub-tropics (10°N to 30°N), land-surface rainfall has decreased on average by 0.3 % per decade. In the recent years, there has been occurance of extreme events like severe droughts experienced recently in Iran, Turkey, Jordan and Pakistan. Jameel (2004) reported that most countries in the Middle East are suffering prolonged water shortages since about 2001.

Snow cover and ice extent: In the Himalayas regions, glaciers are melting rapidly. The Dokriani Barnak glacier retreated more than 20 meters in 1998 alone and in Nepal the Gangotri glacier receded by 600 meters in the last 50 years. In the Tien Shien mountains, China, glacial ice has experienced 25% reduction in the past 40 years. Jameel (2004) reported that ice over Mount Kilimanjaro has decreased from 12 sq. km. in 1910 to about 3 sq. km. in the year 2000.

Key changes at glance

Table 6.1 Key climate changes during 20th century are summarized in table below

Indicator	Change
Atmospheric concentration of carbon dioxide	280 ppm for the period of 1000-1750 to 368 ppm in year 2000 (~30 % rise)
Atmospheric concentration of methane	700 ppb for the period 1000-1750 to 1750 ppb in year 2000 (~ 150 % rise)
Tropospheric concentration of Ozone	Increased by 35 + 15 % from 1750 to 2000; varies with region.
Global mean surface temperature	Increased by 0.6oC over 20th century (30 % uncertainty).
Frequency and severity of droughts	In parts of Asia and Africa, frequency and severity of droughts rose in recent decades
Snow cover	Decreased by ~10 % since 1960s
El Nino Events	More frequent and intense during last 20-30 years as compared to previous 100 years.

Source: Adapted from Jameel (2004)

Emission of Green House Gases (GHGs)

Because of rapid population growth, burining of fossil fuels like wood, coal, oil and natural gases, large amount of carbon monoxide and carbon dioxide gases, as well as nitrogen oxide are emitted to the environment. Increased production of green house gases create an "insulating blanket" in the atmosphere, causing the temperatures to warm up and this would cause the polar icecaps to melt, resulting in a rising ocean level and the flooding of many coastal areas, including large cities of the world. N_2O emission from agriculture is projected to increase by 35-60 % in 2030 due to increased use of nitrogen fertilizer use and animal manure (FAO, 2003).

6.3   Adverse Effects of Climate Change on Soil Properties

Effects of higher temperature: Along with plant growth, moisture and water availability in the soil will be affected by increase in atmospheric temperature, irrespective of any change in precipitation. Higher evaporation rate in higher temperature condition will reduce the level of plant available soil moisture. Higher evaporation will aggravate the problem of soil salinity in low rainfall area. Reduced moisture availability would accelerate the existing problems of infertile soils, soil erosion and poor crop yields. In the extreme case, a reduction in moisture will lead to desertification. Climate change will modify rainfall, evaporation, runoff, and soil moisture storage. Increased evaporation from the soil and accelerated transpiration in the plants themselves will cause moisture stress; as a result there will be a need to develop crop varieties with greater drought tolerance. The irrigation demand is projected to increase in a warmer climate, bringing increased competition between agriculture-already the largest consumer of water resources in semiarid regions-and urban as well as industrial users. Falling water tables and the resulting increase in the energy needed to pump water will make the practice of irrigation more expensive, particularly when with drier conditions more water will be required per acre. Additional investment for dams, reservoirs, canals, wells, pumps, and piping may be needed to develop irrigation networks in new locations.

Finally, intensified evaporation will increase the hazard of salt accumulation in the soil. Link et al. (2003) observed that soil warming and drying led to a 32% reduction in soil C over a five year time period, a much more rapid reduction in soil C than reductions that have been observed due to increased tillage.

Soil fertility and erosion: Higher air temperatures cause increase in soil temperature if soil moisture is not too variable. Warmer soil will enhance the natural decomposition of organic matter and increase the rates of other soil processes that affect fertility. More amount of fertilizer will be required to counteract these processes and to take advantage of increased atmospheric CO_2 for the potential crop growth. The natural cycles of plant nutrients like carbon, nitrogen, phosphorus, potassium, and sulfur-in the soil-plant-atmosphere system is likely to accelerate in warmer soil conditions, thus enhancing CO_2 and N_2O and other GHG emissions. Nitrogen becomes plant available in a biologically usable form through the bacterial nitrogen fixation in the soil. This biological nitrogen fixation causes greater root development, is also predicted to increase in warmer conditions and with higher CO_2, under non- limiting soil moisture conditions. Drier soil conditions will suppress both root growth and decomposition of organic matter, lesser stable aggregate formation and will increase single grain soil structure and thus enhancing the soil's vulnerability to wind erosion, especially if winds intensify. High intensity rainfall will increase soil and water erosion.

Carbon and nitrogen are two major important components of soil organic matter (Brady and Weil, 2008). Organic matter is important for soil physical, chemical and biological properties, including structure formation and maintenance, water holding capacity, cation exchange capacity, and for the supply of nutrients to the soil ecosystem. Soils with an adequate amount of organic matter are more fertile than those soils depleted in organic matter. Soils naturally sequester C through the soil-plant system as plants photosynthesize and then add dead tissues to the soil (Lal, 1998). Depending on the oxygen status of soil system, the microbial Carbon is naturally emitted from soils as CO_2 (from well aerated soil) and CH_4 (from anaerobic soils) gases through microbial respiration. The balance between C added to the soil and C emitted from the soil determines the overall loss or gain of soil carbon in a particular soil (Mosier, 1998 and Schlesinger, 1995). Increase of soil C indicates carbon sequestration from the atmosphere. Decrease in soil C indicates release of C from soil through CO_2 and CH_4.

Anthropogenic activities strongly influence the C balance in managed field. Conservation agriculture (CA) practices like no-till systems may result in lower CO_2 emissions from and greater C sequestration in the soil as compared to conventional tillage system (Figure 1) (Post., 2004, Hobbs et al., 2010). Dendooven et al. (2012) reported that CA practices enhance soil organic matter (SOM) content and nutrient availability by incorporating green manure/ cover crops (GMCC's) in the system and returning their residues to soil (Table) Some studies have reported that no-till systems may accumulate higher C in the upper 15–20 cm of the soil with no increase in C when the entire soil profile is considered (Bakker *et al.*, 2007 ; Blanco-Canqui and Lal, 2008).

Table 6.2 The amount of organic C (Mg ha^{-1}) under zero-till, residue retention and wheat–maize crop rotation (CA), and conventional tillage with crop removal and maize monoculture (CT) between 1996 and 2006 (Dendooven et al., 2012).

Year	CT (Mg ha^{-1})	CA (Mg ha^{-1})
1996	1.80	8.87
1997	2.71	7.83
1998	2.36	10.83
1999	0.48	6.54
2000	2.34	10.40
2001	0.85	8.86
2002	2.35	11.83
2003	1.16	9.74
2004	1.43	10.27
2005	2.48	10.24
2006	2.30	9.38

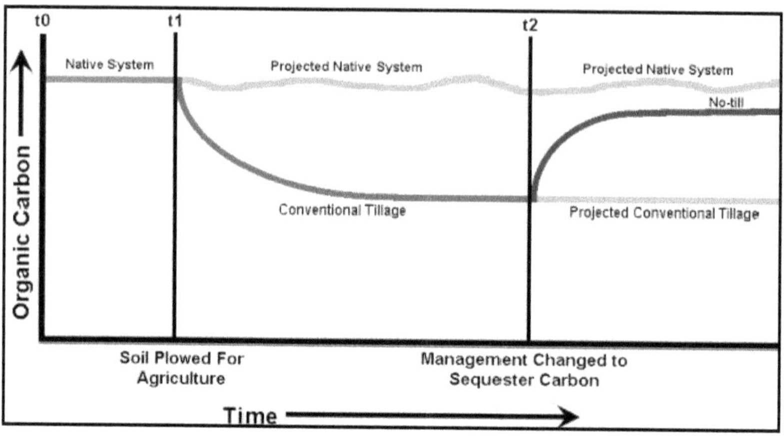

Fig. 6.1 Tilling a native soil leads to reduced soil organic C levels, while management changes such as a conversion to no-till techniques may lead to increased soil organic C as compared to conventional tillage techniques (Brevik, 2012).

Management decisions can restrict the ability of a soil to sequester C as well. For example, the extensive use of heavy equipment in modern production agriculture has made soil compaction a major problem that has been shown to limit C sequestration [28–30]. Organic soils can be a particular C management challenge as they typically form in wet conditions and have to be drained for agricultural uses. This drainage changes the soil environment from anaerobic to aerobic, which speeds decomposition of the organic matter in the soil and releases greenhouse gases into the atmosphere. Methane is another part of the C cycle associated with soils. Agriculture accounts for about 47% of annual global anthropogenic emissions of CH4 (Smith *et al.,* 2007]; the main anthropogenic source of soil-derived methane is rice (*Oryza sativa*) production (Stepniewski *et al.,* 2011). Hu et al. (2001) concluded that different physical properties of soils controlled by the dominant vegetation leads to differences in methanotroph survival and gas exchange rates. Aerobic rice cultivation and other means of reducing the period of soil flooding during rice cultivation leads to less CH_4 production (Stepniewski et al., 2011). Wassmann et al. (1993) found that addition of mineral based K – fertilizer had no effect on CH_4 production in rice soils whereas P fertilizers decreased CH_4 emissions (Lu *et al.,* 2011).

Nitrous oxide (N_2O) is produced when soil water contents is around field capacity and biological reactions in the soil convert NO^{3-} to NO, N_2O, or N_2 (Mullen, 2011). Emissions of N_2O are usually lower in organic farming systems than in conventional systems (Stepniewski *et al.*, 2011).

6.4 Adaptations Strategies for Mitigating Adverse Effects of Climate Changes

The adaptation of agriculture to moderate climate change can be facilitated by improving irrigation, developing less water-demanding and more heat-resistant crop varieties, using minimum tillage and other practices to improve nutrient and moisture retention in soil, and changing timing of planting/harvesting and other management activities. Any strategy adopted to increase crop production will mitigate the adverse effects of climate changes. The most adverse effects of climate changes related to soil problems are high temperatures and extreme events of drought and floods. High temperatures are the result of emission of Green House Gases (GHGs). This effect can be mitigated partly by following soil management strategies.

6.4.1 Integrated Plant Nutrient Management (IPNM) for Different Crops

Variable rate fertilizer management (precision agriculture)

Use of nitrification inhibitors

Incorporation of crop residues

Management of marginal lands (salt affected soils) restoration of crop productivity of marginal lands

Moisture Conservation

These manage strategies will ultimately mitigate the emission of GHGs into the air and improve the sequestration of CO_2.

Conservation agriculture for climate smart agriculture: Conservation farming is now recognised globally as the most important integrated farming system with the potential to reduce the impacts of agriculture, improve and protect the natural resource base, address carbon emissions and climate change issues and improve social and economic outcomes for farming communities all over the world. The global concern about soil degradation is helping to support policies towards conservation farming at the international level. The link between carbon sequestration in soil, global warming and the role of conservation farming is now recognised by agricultural policy makers worldwide. Conservation Agriculture can assist in the adaptation to climate change.

By improving the resilience of agricultural cropping systems and hence by making them less vulnerable to extreme climatic situations.

Improved soil structure and high water infiltration rate reduce the chances of flooding and erosion after high intensity rainfall.

Increased soil organic matter improves soil water holding capacity and helps to sustain crops in drought periods.

Yield variations under CA in extreme years are less pronounced than conventional agriculture.

CA mitigates climate change by emitting less amount of green house gases like carbon dioxide, nitrous oxide and (through reduction in fossil fuel consumption and elimination of crop residue burning) and sequestrating soil organic carbon over a long period.

In paddy cultivation, no-till systems and adequate water management minimizes the release of greenhouse gases like methane and nitrous oxides.

References

Bakker, J.M.; Ochsner, T.E., Venterea, R.T. and Griffis, T.J (2007). Tillage and Soil Carbon Sequestration-What do We Really Know? *Agric. Ecosyst. Environ., 118,* 1–5p.

Blanco-Canqui, H., Lal, R. (2008). No-tillage and Soil-profile Carbon Sequestration: An on-farm Assessment. *Soil Sci. Soc. Am. J., 72,* 693–701p.

Brady, N.C., Weil, R.R. (2008). The Nature and Properties of Soils, 14th ed.; Pearson Prentice Hall: Upper Saddle River, NJ, USA.

Brevik, E.C. (2012). Soils and Climate Change: Gas Fluxes and Soil Processes. *Soil Horiz.*, *53*, doi:10.2136/sh12-04-0012.

FAO. (2003). World Agriculture: Towards 2015/2030. An FAO Perspective. Food and Agriculture Organisation, Rome, Italy. 97p.

Hobbs, P.R., Govaerts, B. (2010). How Conservation Agriculture can Contribute to Buffering Climate Change. In *Climate Change and Crop Production*; Reynolds, M.P., Ed.; CPI Antony Rowe: Chippenham, UK. 177–199p.

Hu, R., Kusa, K., Hatano, R (2001). Soil Respiration and Methane Flux in Adjacent Forest, Grassland, and Cornfield Soils in Hokkaido, Japan. *Soil Sci. Plant Nutr*, *47*, 621–627p.

IPCC. Summary for Policymakers. In *Climate Change 2007: The Physical Science Basis*; Contribution of Working Group I to the Fourth Assessment Report of the Intergovernmental Panel on Climate Change; Solomon, S., Qin, D., Manning, M., Chen, Z., Marquis, M., Averyt, K.B., Tignor, M., Miller, H.L., Eds.; Cambridge University Press: Cambridge, UK, 2007; 1–18p.

Jameel, M. (2004). Climate Change: Global and OIC Perspective. COMSTECH Islamabad. 61p.

Lal, R.; Kimble, J.; Follett, R.F. (1998). Pedospheric Processes and the Carbon Cycle. In *Soil Processes and the Carbon Cycle*; Lal, R., Kimble, J.M., Follett, R.F., Stewart, B.A., Eds.; CRC Press: Boca Raton, FL, USA, 1–8p.

Link, S.O.; Smith, J.L.; Halverson, J.J.; Bolton, H., Jr. (2003). A Reciprocal Transplant Experiment within a Climatic Gradient in A Semiarid Shrub-Steppe Ecosystem: Effects on Bunchgrass Growth and Reproduction, Soil Carbon, and Soil Nitrogen. *Glob. Change Biol.*, *9*, 1097–1105.

Luc Dendooven, Vicente F. Gutiérrez-Oliva, Leonardo Patiño-Zúñiga, Daniel A. Ramírez-Villanueva, Nele Verhulst, Marco Luna-Guido, Rodolfo Marsch, Joaquín Montes-Molina, Federico A. Gutiérrez-Miceli, Soledad Vásquez-Murrieta, Bram Govaerts (2012). Greenhouse Gas Emissions under Conservation Agriculture Compared to Traditional Cultivation of Maize in the Central Highlands of Mexico. *Science of the Total Environment*, 431: 237–244.

Mosier, A.R. (1998). Soil Processes and Global Change. *Biol. Fertil. Soils*, *27*, 221–229.

Mullen, R.W. (2011). Nutrient Cycling in Soils: Nitrogen. In *Soil Management: Building a Stable Base for Agriculture*; Hatfield, J.L., Sauer, T.J., Eds.; Soil Science Society of America: Madison, WI, USA, 67–78p.

Post, W.M.; Izaurralde, R.C.; Jastrow, J.D.; McCarl, B.A.; Amonette, J.E.; Bailey, V.L.; Jardine, P.M.; West, T.O.; Zhou, J. (2004). Enhancement of Carbon Sequestration in US Soils. *Bio-science*, 54, 895–908p.

Schlesinger, W.H. (1995) An Overview of the Carbon Cycle. In *Soils and Global Change*; Lal, R., Kimble, J., Levine, E., Stewart, B.A., Eds.; CRC Press: Boca Raton, FL, USA, 9–25p.

Smith, P.; Martino, D.; Cai, Z.; Gwary, D.; Janzen, H.; Kumar, P.; McCarl, B.; Ogle, S.; O'Mara, F.; Rice, C.; et al. (2007). Agriculture. In *Climate change 2007: Mitigation*; Contribution of Working Group III to the Fourth Assessment Report of the Intergovernmental Panel on Climate Change; Metz, B., Davidson, O.R., Bosch, P.R., Dave, R., Meyer, L.A., Eds.; Cambridge University Press: Cambridge, UK, 497–540p.

Stepniewski, W.; Stepniewski, Z.; Rożej, A. (2011). Gas Exchange in Soils. In *Soil Management: Building a Stable Base for Agriculture*; Hatfield, J.L., Sauer, T.J., Eds.; Soil Science Society of America: Madison, WI, USA, 117–144p.

Wassmann, R.; Schütz, H.; Papen, H.; Rennenberg, H.; Seiler, W.; Aiguo, D.; Renxing, S.; Xingjian, S.; Mingxing, W. (1993) Quantification of methane emissions from Chinese rice fields (Zhejiang Province) as influenced by fertilizer treatment. *Biogeochemistry*, 20, 83–101p.

7

Climate Change and Drought Governance in Dry Land Agriculture

Devi Prasad Juvvadi
Centre for Good Governance, Hyderabad.

7.1 Introduction

In recent years, concern has grown worldwide that droughts may be increasing in frequency, severity, and duration due to changing climatic conditions. This is substantiated by increase in extreme climate events documented (Sivakumar, 2012 and Peterson et al., 2013). Responses to drought by governments throughout the world have generally been reactive, poorly coordinated and untimely and have been typically characterized as "crisis management" (Wilhite and Pulwarty, 2005).

The provision of relief or assistance to the drought affected areas in many countries has been shown to increase vulnerability to future drought episodes by reducing self-reliance and increasing dependence on central or federal government for drought relief assistance. Hence, it is imperative that governance of drought should include emergency relief the affected regions as a safety net for farmers who are most vulnerable and at the same time promote self-reliance through technologies for agricultural risk management.

7.2 Climate Change

Climate change is recognized as a major factor and important driver of ecological changes in dry land areas. More frequent and intense extreme weather events in many dry land regions are projected to increase and considered to be a major consequence of global climate change. This will result in a significant increase in the extent of dry land and these are also likely to be affected by more intense and longer drought periods (IPCC, 2007). Several projections in climate change trends for dry lands indicate an upward trend in temperature leading to higher levels of evaporation from soils, crops, and water bodies, while consequently adding stress to crops and animal health. It is projected that climate change will lead to a decrease in availability and quality of water (10% -30% in the next 40 years), while extreme weather events such as droughts and floods will increase in number and/or intensity (Davies et al.2012). Additionally, IPCC projections also show that majority of rainfed areas are likely to receive less rain (Venkateswarlu et al., 2009).

Increasing temperatures are expected to add to water management problems by increasing water stress through additional loss of moisture from the soil.It has also been estimated that by 2020, about 75 to 250 million people are likely to be exposed to increased water stress. In the scenario of continuation of the existing production practices, rain-fed agricultural yields could result in a reduction by up to 50% (IPCC, 2007). Climate change will therefore result in reduced agricultural productivity in dry lands and is expected to have severe impacts on food security (Davies et al., 2012) adding pressures to an already fragile water situation and the further undermining of livelihoods. It is also expected that the projected higher frequency of dry spells might encourage dry land farmers to increase water withdrawals for irrigation.

7.3   India Scenario

Climate change projections for India indicate an overall increase in temperature by 2-4 °C by 2100 with no substantial change in precipitation quantity (Kavikumar 2011). Studies also indicate that India is likely to be affected with one of the highest agricultural productivity losses in the world in accordance with the observed and projected climate change patterns. Most of the studies projected that the decreased yield in rainfed and dry land wheat and rice and loss in farm net revenue between 9 to 25 per cent for a temperature increase of 2 to 3.5 °C. Researchers (Sinha and

Swaminathan, 1991) projected that an increase of 2 °Celsius in temperature could decrease rice yield by about 0.75 tons/ha; and a 0.5 ° C increase in winter temperature could reduce wheat yield by 0.45 ton/ha in India. Similarly, it has also been estimated that for every 1 °C rise in temperature the corresponding decline in rice yield would be about 6 % (Saseendran et al., 2000). The impacts of climate change in drylands are therefore likely to lead to even more people and larger areas of land being affected by water scarcity and an accompanying increase in risk of declining crop yields of small and marginal farmers who are least able to adapt due to poverty.

7.4   The Drought Conundrum

The impacts of droughts in dry lands have drastically increased as a result of increased frequency, severity and duration of droughts as well as the narrowing of the gap between supply and demand of water (Wilhite, 2014). The increase in droughts will have exponential impacts on agriculture, but despite this the issue of formulation and adoption of drought policies seems to have not been adequately addressed. This therefore necessitates proactive action at local as well as national levels.

A national level drought policy approach is required in order to focus holistically on comprehensive monitoring, early warning and information systems, impact assessment procedures, risk management measures, drought preparedness plans, and emergency response programs. It is also the need of the hour to to avoid responding to drought in a reactive, crisis management mode.

Drought is contrasted with aridity since temporally speaking, drought is of a temporary duration unlike aridity, which is characterized as a somewhat permanent feature of the climate. In the same vein, seasonal aridity (i.e., a well-defined dry season) also needs to be distinguished from drought. There is ambiguity surrounding the differentiation of the terms aridity and drought among scientists and policy makers. It is important to note that drought also occurs in high and rainfall areas where different management practices are required.

Wilhite et al., (2014) explained in detail about drought being a slow onset natural hazard whose effects accumulate slowly over a substantial period of time. Therefore due to this operational-ization, the onset and the end of drought becomes difficult to determine with scientists and policy makers disagreeing on the declaration of an end to drought. Further, the absence of a precise and universally accepted definition confounds issues as to whether droughts exist or not. The resolution of the degree of severity is another issue altogether. To resolve this, drought may be considered in relative terms in contrast to absolute terms and that the definitions of drought must be region and application (or impact) specific (Wilhite, 2014). It is in this light that a precise, universal definition of drought is of little value.

This is also especially significant considering that the impacts of drought are not structural in nature. Moreover, the impacts of drought are generally much larger than impacts occurring from other natural hazards such as storms, earthquakes, floods etc. The impacts are also spread over a much larger area in geographical space terms. This unique feature of droughts also therefore contributes to the problem of the quantification of impact assessment and the social, economic and environmental risks associated with it thereof, since the impacts of droughts continue to linger in the economy as well as the environment for a much larger period of time usually for years and also sometimes decades. As a consequence, the developments of impact assessments have been hindered, and this has also reflected in the way drought preparedness plans have been formulated.

7.5   Effect of Drought on Agriculture

In India, the El Niño-Southern Oscillation (ENSO) related droughts have been found to be contributing to periodic declines in agricultural output. Drought directly contributes to fall in crop production because of inadequate and poorly distributed rainfall resulting in water stress during crop growth and due to. The water stress periods are called as breaks in the monsoon and are found to be frequent in monsoon months, particularly in August occurring every two and half years in Telangana.

The dry spells, could be classified as one of early season drought, midseason drought and terminal drought, and are found to be lasting on average for a few days or some times more which result in the partial or total failure of crops.

A brief explanation of the droughts is offered here (Sharma et al., 2006):

Type	Nature	Effect
Early Season Drought	Occurrence due to delayed onset or prolonged dry spell	Results in seedling mortality with crops suffering from acute water shortage of water during reproductive stage
Mid season drought	Occurrence due to inadequate soil moisture availability	Results in stunted growth if it occurs at vegetative phase; will result in an adverse effect on crop yield if it occurs at flowering or early reproductive stage.
Late season or terminal drought	Occurrence as a result of early cessation of monsoon rains. Further, these conditions are often associated with late commencement or weak monsoon activity.	Results in an increase in ambient temperatures leading to forced maturity.

Thomas et al. (2007) through a comprehensive analysis has summarized the key impacts of climate change and climate variability resulting in drought and they include:

- Reduction in crop yields and agricultural productivity with subsequent threats to the food security of dryland regions.

- More erratic rainfall patterns and difficulties in determining timings of sowing and harvesting, and the selection of suitable crops with varying durations.

- Reduced availability of water in already water scarce regions coupled with extreme rainfall events with increased loss of water via run off, etc.

- Complete loss of crops resulting from extreme events such as prolonged droughts and torrential rains.

- Slow pervasive loss of soil fertility through loss of soil carbon from erosion and higher decomposition of soil organic matter as a result of higher temperatures, reduced soil moisture and moisture storage capacity.

- Lower livestock productivity from heat dissipation and reduced availability of feed and fodder.

- Alterations in pest and disease risks for both crops and animals (and humans) as temperatures increase.

- Changes in agro-ecologies and the threats from new invasive plant and animal species.

- Reduction of biodiversity of key crop species through habitat change and loss.

Thomas (2007) further suggests that the impacts will result in increased poverty, and reduced livelihood opportunities that subsequently results in increased rates of migration. From this it could be inferred that there is a pressing need to enhance the practices of natural resource management in order to prepare the famers to adequately respond to climate change in dryland areas.

7.6   Governance of Drought in Climate Change Arena

It is important that both policy formulators and agricultural scientists should address the drought in rainfed agriculture or dryland farming from different viewpoints and evolve appropriate technologies in order to improve incomes of small and marginal farmers who are more prone to climate change and resulting drought. Hence governance of drought is discussed as planning by policy makers, strategies, technologies and practices for management of drought by researchers.

7.7   Drought Planning

Drought planning could take the shape of response planning or mitigation planning and can be defined as "actions taken by government or individual farmers, and others before drought occurs with the purpose of reducing or mitigating impacts and conflicts arising from drought" (Wilhite, 2014). Since drought due to climate change is more frequest in dryland agriculture, policy makers should anticipate drought and prepare drought plans. A comprehensive drought planning mechanism has been describd by Wilhite (2014) which has been implemented in different countries with suitable modifications. The 10 step process is given here:

1. ***Appoint a drought task force:*** Based on the deficient rainfall or deficient sowings due to poor monsoon, the government initiates the drought planning process through appointment of a drought task force. The task force should reflect the multi disciplinary nature of drought and its impacts and it should include appropriate representatives of government agencies (state and central) and universities where appropriate expertise is available.

2. ***State the purpose and objectives of the drought mitigation plan:*** As its first official action, the drought task force should state the general purpose for the drought preparedness plan. Government officials should consider many questions as they define the purpose of the plan, such as the following:
   - Purpose and role of government in drought mitigation and response efforts
   - Scope of the plan
   - Most drought –prone areas of the state
   - Historical impacts of drought
   - Historical response to drought;

3. ***Seek stakeholder participation and resolve conflict:*** The task force members must identify all citizen groups (stakeholders) that have a stake in drought planning and their interests. These groups must be involved early and continuously for fair representation and effective drought management and planning. Discussing concerns early in the process gives participants a chance to develop an understanding of each other's viewpoints, and to generate collaborative solutions.

4. *Inventory resources and identify groups at risk:* An inventory of natural, biological, and human resources, including the identification of constraints that may impede the planning process, may need to be obtained by the taskforce from local departments and agencies. These constraints may be physical, financial, legal, or political.

5. *Prepare and write drought plan:* This step describes the process of establishing relevant committees to develop and write the drought preparedness plan. The plan should have three primary components: monitoring, early warning and information delivery coupled with prediction; risk and impact assessment; and mitigation and response.

   The steps in this process include the following:
   - Identify impacts of recent and historical droughts.
   - Identify drought impact trends.
   - Prioritize impacts.
   - Identify mitigation actions that could reduce short- and long- term impacts.
   - Identify triggers to phase in and phase out actions during drought onset and termination.
   - Identify agencies and organizations to develop and implement actions.

6. *Identify research needs and fill institutional gaps:* As research needs and gaps in institutional responsibility become apparent during drought planning, the drought taskforce should compile a list of those deficiencies and recommend possible remedies to the government.

7. *Integrate science and policy:* An essential aspect of the planning process is integrating the science and policy of drought management. The policymaker's understanding of the scientific issues and technical constraints involved in addressing problems associated with drought is often limited. Likewise, scientists generally have a poor understanding of existing policy constraints for responding to the impacts of drought. The communication and

understanding between the science and policy communities must be enhanced if the planning process is to be successful.

8. *Publicize the drought mitigation plan, build public awareness and consensus:* If there has been good communication with the public through- out the process of establishing a drought plan, citizens may already have better-than-normal awareness of drought and drought planning by the time the plan is actually written. This should include writing news stories on how the drought plan is expected to relieve impacts of drought in both the short and long term.

9. *Develop education programs:* A broad-based education program with folk songs and kalajathas to raise awareness of short- and long-term water supply issues will help ensure that people know how to respond to drought when it occurs and that drought planning does not lose ground during non-drought years.

10. *Evaluate and revise drought mitigation plan:* Periodic testing, evaluation and updating of the drought plan are essential to keep the plan responsive to the needs of the state and its citizens. To maximize the effectiveness of the system, two modes of evaluation must be in place; Ongoing evaluation and Post-drought evaluation.

Response to drought in many countries has traditionally assumed a crisis management centric approach, which has been considered to be both ineffective (i.e., assistance poorly targeted to specific impacts or population groups), poorly coordinated and untimely the approach also been criticized for having done little to reduce the risks associated with drought.

The above framework therefore provides a process overview for the development of state as well as national level drought policies and preparedness plans that may help in improving drought preparedness levels and measures.

7.8 Drought Management Strategies

In the context of climate change and variability resulting in drought, governance of dryland agriculture need to opt for resilience of agricultural systems and adaptive capacities. Farmers must be encouraged to adopt proven technologies and practices to climate induced problems in agriculture. Innovative practices, processes, products, decision tools and livelihood options appropriate to the farming situations must be encouraged to tackle climate risks in agriculture.

Indian Meteorological Department (IMD) cautions about weak monsoons and this helps to develop long and short term drought management strategies. With increase in frequency and intensity of droughts due to the El Nino effect during the last two decades, government's efforts of increasing agricultural production have suffered. Long term strategies need to be evolved to tackle the challenge of drought and also to improve climate change based agricultural vulnerabilities arising from droughts. As many Indian states are increasingly becoming highly vulnerable from a drought perspective, an appropriate drought management strategy is needed to prepare farmers to adapt and respond to climate change. A drought perspective is also essential in promoting a sustainable increase in food grain production and in building the necessary resilience and adaptive capacity, while at the same time helping reduce green house emissions associated with agriculture.

Technology may also be leveraged by IMD and ICAR institutes in providing timely weather updates that have a high degree of localization. Specific apps or SMS alerts maybe disseminated to the target audience comprising of farmers, extension functionaries and other stakeholders. An example for this could be the Kisan SMS portal that disseminates weather related information. To pursue this appropriate capacity building strategies may also be evolved to implement climate-smart technologies and effective use of weather advisories. Knowledge management and capacity building at the grass-root level therefore becomes a basic necessity for preparing for climate change eventualities. Insurance models should also be properly calibrated. One method for instance is to tie it up with weather and mitigate the risk by providing adequate insurance to farmers afflicted with droughts.

Joshi & Aggarwal (2014) suggested a range of strategies and measures that will help in averting the negative impact of drought and also bolster national policies on agriculture. Technology plays a crucial role in helping productivity in different conditions of aridity, particularly in dryland areas and those characterized by droughts. The technological intervention strategies maybe carried out at village level to improve resource use efficiency, increase farm production and income, and minimise climatic risks. Some of the strategies are as follows:

Selection of Drought-Resistant Varieties

It is important that varieties which have proven genetic character to withstand longer periods of drought are chosen so that the crops can do well even in situations where the intervals between rainy days are long.

Early Maturing Varieties

Where the distribution and the amount of rain is unpredictable, it is important to select varieties which have a shorter duration life cycle (seed to seed) to cut down the water requirements of the crop. In drought-prone areas, the success rate of short duration crops is greater than long-duration crops.

Role of Vegetation

Vegetation is crucial in preserving productive soil and conserving rain water for sustaining life. Soil and water need to be preserved for crop production (both annual and plantation crops) as well as plants that give fuel, fodder, fruits, industrial raw materials, medicinal and aromatic plants and the like.

Seed Rates

In normal season, swing is done with the normal seed rate. However, if there is a drought during the plant's growth period and wilting is likely to occur, selective thinning is recommended to reduce the plant population to effectively use the scarce soil moisture among fewer plants. In late season, where the monsoon is moderately delayed, normal cropping with reduced seed rate is advised.

Wider Spacing

In all drought-prone areas, the most important objective is to raise a successful crop under scarce soil moisture conditions. One of the recommended practices is wider spacing between rows and between plants within the row. This reduces plant population and competition between plants for scarce soil moisture. Fewer plants have greater access to limited available soil moisture.

Weeding

Frequent weeding is an important part of dryland agriculture. Line sowing and mechanical weeding, with appropriate size of blade harrows, remove unwanted vegetation which competes with the main crop. It is not uncommon to see the dryland farmer hitching several blade harrows to one yoke and a pair of bullocks. Weeding within rows can be done using hand hoes. Removal of unwanted vegetation helps the main crop obtain greater accessibility to soil moisture and plant nutrients for its own growth.

Mixed/Inter Cropping

Mixed cropping of different crops along with the main crops, such as millets and different legumes, is an insurance against the vagaries of the monsoon. The different root systems of mixed crop feed at different depths of the soil. Moreover, mixing cropping provides small quantities of grain of different kinds for home consumption at different times.

Mulching

Mulches are ground covers that prevent the soil from being washed away, reduce evaporation, increase infiltration, and control growth of unwanted weeds. Mulch can be organic crop residue, pebbles, or materials such as polythene sheets. Mulching prevents the formation of hard crust after each rain. Organic mulches add plant nutrients to soil upon decomposition. Use of blade harrows between rows also creates "dust mulch" by breaking the continuity of capillary tubes of soil moisture.

Contingent Planning

With every care taken to undertake timely agricultural operations, it is still possible that the whole operation becomes a gamble due to unpredictable monsoons. The main crop could fall in the early part of its life cycle. In such cases, the farmer should come up with an alternate crop that can mature in a very short time and under hard conditions to take advantage of what is left of the rainy season. Contingent planning helps catch and make the best use of late rains. Advance planning is necessary in selecting a contingent crop. And all the requisites for its sowing should be ready within the main season itself. Credit for farmers must be made available at the right time.

Rainwater conservation

Soil and water conservation measures consist of agronomical and mechanical methods. Agronomic methods to be supported with mechanical measures are to be developed where land slope exceeds permissible limits and runoff gains erosive velocities.

Efficient irrigation systems and practices

One of the great challenges for dryland agriculture is establishing plants and growing crops on sites that are seasonally dry or dry all year. This is becoming increasingly important as climate change creates new areas of drought and more erratic rainfall patterns. Drip systems are well but low cost simple irrigation methods that would help to increase yields have to be used. The efficient traditional methods of irrigation include: deep pipes, buried clay pots, porous capsules, wicks, porous hose, and sub irrigation with perforated pipe. These require much less water and work well on slopes. They also reduce weed growth dramatically and ensure that water is used to grow the crop, not weeds.

Soil test based nutrient management and balanced fertilisation

The soil test based nutrient management or balanced fertilisation implies meeting the individual nutrient needs of crops according to their physiological requirements and expected yields. This means the deliberate application of all nutrients that the soil cannot supply in adequate amounts for optimum crop yields. It depends on soil test values and requires estimates of what crops remove. There is no fixed recipe as it is soil and crop-specific.

It is hoped that these strategies with policy initiatives and support could help in tackling the problem of drought.

7.9 Drought Management Practices

7.9.1 Conservation Agriculture

Small and marginal farmers face many challenges as a result of the current agricultural cultivation practices. These challenges could be considered endemic to the dryland agricultural systems and in order to address such problems facing dryland agriculture systems, conservation agriculture has evolved as a concept for addressing specific problems faced by smallholder farming systems in the tropics (Hobbs 2007; Hobbs et al. 2008) This concept is gaining popularity in the tropics and is being promoted by many national and international organizations. Originally advocated by the FAO (2008), conservation agriculture is a concept for resource-efficient agricultural crop production based on an integrated management of soil, water and biological resources combined with external inputs. It is based on three principles that enhance biological processes above and below the ground (1) minimum or no mechanical soil disturbance; (2) permanent organic soil cover (consisting of a growing crop or a dead mulch of crop residues); and (3) diversified crop rotations.

7.9.2 Managing Water Resources

Effective and efficient use of water resources is essential to overcome the climate extremes. It is also central to unlocking the development potential of dryland areas (Nkonya et al. 2011). Over the years, many approaches have been followed for water resources management and development in dryland areas. These include development of groundwater resources for domestic and productive uses, capturing more surface water in the soil, soil and water conservation, and water harvesting. In the rainfed areas, rainwater is the main source of irrigation . Groundwater is another important source for irrigated agriculture as it generally furnishes reliable and flexible inputs of water. However, it has been established that it's current use efficiency in crop production is low (30-45%). This establishes that groundwater is instrumental in managing risk and optimizing food production in the rainfed areas.

Depleting groundwater is a serious problem in dryland regions and in promoting integrated watershed development as an effective strategy for sustainable development of dry land areas. integrated watershed can become the growth engine for sustainable development of dry land areas by improving the performance of 2/3rd watersheds in the country (Wani et al. 2009). In most of the developed watersheds with concerted efforts to manage rainwater, the groundwater availability is improved not only in the watershed, but also in the downstream areas due to increased groundwater recharge (Pathak et al. 2007). These interventions have resulted in increased surface and groundwater availability with accompanying private and public investments and have contributed to substantial increase in incomes as well as improved livelihoods (Pathak et al., 2007).

7.9.3   Genetic Resources and Adoption of Climate Resilient Crops

It has been found that in many instances, farmers dependent on rainfed farming adopt intercropping and mixed cropping strategies for crop diversification in regions that have limited rainfall and growing season. Other strategies practiced in the dryland areas include early planting and harvesting at physiological maturity of crops, less number of tillage operations, deep placement of fertilizers (Parida, 2010).

Attention on a priority basis needs to be paid to agricultural biodiversity and crop germplasm exploration since these are of relevance to the dryland and water stressed environments requires priority attention in the coming days. Similarly crop breeding research must also be fostered. Seeds and plants that exhibit tolerance to enhanced temperature, reduced precipitation and other atmospheric stresses caused by climate change need to collected and conserved for the purpose. Evaluation of crop germplasm including wild relatives, land races, extant varieties, and modern varieties and breeding stocks could help in unravelling traits than could prove more useful since it will aid in the conservation of the rich genetic diversity of resources.

The natural variability could also be conserved and evolved as a result through continued efforts of the farming communities over the years. This could prove to be the most important and cost effective aspect in allowing agriculture to adapt to the various dimensions of climate change (Dawson et al., 2011).

Significant progress had been made during the last two decades in the areas of drought tolerance in model plant systems, and by comparative genomics in crop plants (Jung and Müller 2009). It is expected that the research would provide further impetus to the identification and development of location specific varieties and cultivars those could profitably be used in farming systems in the dryland areas.

7.9.4   Integrated Farming Systems

It is of the view that decreasing land size holding in Indian context poses a serious challenge to the stability, sustainability, productivity and profitability of farming systems (Parida, 2012). A decline in per capital availability of land holdings from 0.5 ha in 1950-51 to 0.15 ha in 2000-01 has been observed and is projected to further decline to less than 0.1 ha by 2020. In light of this, it becomes essential to develop strategies and agricultural technologies that enable adequate employment and income generation, specifically for the small and marginal farmers who constitute more than 85% of the farming community, in the country. It also becomes crucial to shift from crop and cropping system based research to research centred on farming systems for an effective management of available resources by small farmers. Under the gradual shrinking of land holding, it is necessary to integrate land based enterprises like fishery, poultry, duckery, and horticultural crops, etc., within the biophysical and socio-economic environment of the farmers to make farming more profitable and dependable. Farming systems approach, therefore, is a valuable approach to addressing the problems of sustainable economic growth for small and marginal farmers in India since no single farm enterprise is likely to be able to sustain the small and marginal farmers without resorting to integrated farming systems (IFS) for the generation of adequate income and gainful employment through out the year.

In light of improving productivity of the farming system, Parida (2012) also reports that the farming system mode offers unique opportunities such as:

(i) *in situ* recycling of organic residues including farm wastes generated at the farm to reduce the dependency on external inputs,

(ii) decrease in cost of cultivation through enhance input use efficiency as well as engagement of family workforce,

(iii) effective forward and backward linkages within the farm components,

(iv) upgrading of soil and water quality and increased diversity in the fields,

(v) effective water management and productivity,

(vi) nutritional security through soil-plant–animal- human chain.

Integration of various agricultural enterprises viz., cropping, animal husbandry, fishery, forestry etc. in the farming system has great potentialities in agricultural economy since these enterprises supplement farmer incomes and also help in increasing the family labor employment throughout the year (Jayanthi et al. 2002; Singh et al. 1993 and Singh et al. 1997). Research has also suggested that the Integrated Farming Systems (IFS) contributes to higher water use efficiency by about 71% than the conventional system (Channabasavanna et al., 2009) since integrated farming requires less water per unit of production than mono-cropping systems- about 1247 mm of water as opposed to 2370 mm of water for conventional farming system (Jayanthi et al., 2000; Channabasavanna, 2009). In view of this therefore, a well designed integrated intensive farming system based on the traditional agriculture practices that are specific to a particular agro-ecosystem need to be developed, and at the same time aligned with marketing opportunities for the products of small farms, whilst also providing an opportunity for the involvement and engagement of family labour and home consumption of diversified agricultural baskets.

7.10 Recommendations for Major Crops under Climate Change and Vulnerability Scenario

Agronomic Measures for Major Crops in Dry land Agriculture

Murthy (2016) made agronomic recommendations for major crops under climate change variability scenarios for dry land agriculture.

Maize (Zea mays)

- Use recommended high yielding varieties/hybrids/composites
- Ensure optimum plant population of about 66,000/ha
- Well rotten compost or FYM @ 10-12 t/lia shall be incorporated
- Take up weed management up to one month after sowing
- Atrazine 50% WP @ 2 kg/ha in case of light soils and 3 kg/ha in case of heavy soils is to be mixed in 500 L of water and sprayed uniformly within 2-3 days after sowing on moist soil for management of weeds
- Balanced fertilization on the basis of soil testing
- Zinc application in case of its deficiency at 50 kg zinc sulphate/ha
- A total of 400-500 mm of water would be enough for kharif maize, if water losses through different sources are kept to the minimum
- Intercropping of forage maize in grain maize, up to it's knee-high stage
- In sequence crops, maize-groundnut and maize-redgram to be preferred to maize-maize
- Timely plant protection against stemborer and leaf spots
- Zero till drill machine can be used for dibbling the seed. Plant population should be 83,333 plants/ha.

Jowar (Sorghum vulgare)

- Use recommended high yielding varieties/hybrids, seed treatment with Thiram / Captan @ 3g/kg of seed
- Timely sowing before first July to avoid pest problems such as shoot fly and earhead bug
- Seed rate 8-10 kg/ha and spacing 45 x 15 cm
- Proper weed management up to one month after sowing
- Balanced fertilizer management and FYM 10 t/ha
- Flat bed sowing followed by ridging at 30 DAS results in moisture conservation and high yield in red soil
- Sowing with onset of monsoon
- Intercropping Jowar and redgram in 2:1 ratio is more remunerative than sole crop and reduces charcoal not incidence in sorghum and wilt in redgram
- Intercropping of Jowar and cowpea in 2:2 ratio (Jowar in paired rows) enhances the total fodder production without affecting grain yield of Jowar
- Need-based plant protection against insect pests such as shoot-fly, stem borer and earhead bug and also diseases such as sugary disease and grain molds.

Bajra (Pennisetum typhoides)

- Use recommended high yielding varieties/hybrids, seed treatment with salt water (2%) and Thiram @ 3g/kg of seed
- Open pollinated varieties (composites and synthetics)
- Sowing by ridge and furrow method will be effective to conserve moisture
- Intercropping of bajra and redgram in 2:1 ratio gives higher monetary returns
- Maintain weed free field and pre emergence application of weedicide Atrazine @ 4g/Iiter within 48 hours of sowing
- Fertilization @ 60 N, 30 P_2O_5 and 20 K_2O Kg/ha with N in two equal splits - one at sowing/planting and another at about 3-4 weeks after sowing/planting

- Need based plant protection against white ants and root grubs in endemic areas and also against ergot and downy mildew

- Plough the field soon after harvest to bury the ergot inoculums.

Ragi (Elusine Corocana)

- Use recommended high yielding varieties, seed treatment with Thiram, Captan or Carbendazim @ 3g/kg of seed

- Minor land smoothening, before sowing helps in better *insitu* moisture conservation

- Nursery - use 5.0 kg seed in 400 m² area to plant one hectare.

- Weed management in nursery and also up to 3 weeks after planting

- Fertilizer application @ 60 N, 40 P_2O_5 and 30 K_2O kg/ha, with N in two equal splits - one at planting and another 3-4 weeks after planting

- Spraying of carbendazim 0.05% twice - at 50% earhead emergence and also at complete earhead emergence against blast.

Korra (Setaria italica)

- Use recommended high yielding varieties

- Seed treatment with carbendazim @ 2 gm/kg seed

- Weed management up to 30 days after sowing.

- Fertilization @ 40 N and 20 P_2O_5 kg/ha with N in two equal splits - one at sowing and another 3-4 weeks after sowing.

- Intercropping korra + redgram in 5:1.

- In irrigated crop, irrigation should be given after sowing, tillering, ear head emergence, flowering and grain filling stages.

Redgram (Cajarus cajan)

- The integrated pest management of *helicoverpa* 011 redgram summer ploughing, seed treatment with rhizobium culture, following crop rotation and growing intercrops, adjustment of sowing dates with varieties of different durations is recommended

- Release of trichogrammea twice at weekly intervals @ 6500/ha also helps in escaping from helicoverpa
- Adopt IPM against pod borer on redgram
- Intercrop redgram with greengram/blackgram/groundnut sesamum/soybean/sorghum or maize
- Under delayed monsoon conditions, grow redgram as sole crop using high seed rate
- Rabi crops may be intercropped with rabi redgram.

Blackgram (Vigna mungo)

- Do not cultivate blackgram on light soils in uncertain rainfall areas as it is highly sensitive to moisture stress
- Rabi blackgram under ID conditions and rice follows
- Use recommended high yielding varieties, seed treatment with Captan / Thiram / Mancozeb / Carbendazim @ 3g per kg of seed
- Maintain optimum plant population of 30-35 plants/sq.mt
- Spray 1.5-2.0% urea twice at 40 DAS and 50 DAS
- Follow IPM for the management of tobacco caterpillar
- Timely pest and disease management
- Spray twice Mancozeb (0.3%) or Copper Oxychloride (0.3%) at 10 days interval to manage foliar fungal diseases
- Spray twice with karathane (0.1%) + Mancozeb (0.3%) or tridemorph twice at weekly intervals at 50-55 DAS to control the rust
- Foliar nutrition of KNO_3 @ 10 g/L in saline soils.

Greengram (Vigna radiata)

- Take up greengram as catch crop before rice planting
- Adopt line sowing and maintain 30-35 plants per sq.mt
- Use recommended high yielding varieties, seed treatment with Captan / Thiram / Mancozeb / Carbendazim @ 3 g per kg of seed

- Apply 20 kg N, 50 kg P_2O_5/ha basally and incorporate
- Spray 1.5-2.0% urea twice at 30 DAS and 40 DAS
- Grow YMV resistant varieties.
- Light irrigations are always beneficial.

Bengal gram (Cicer arietinum)

- Use recommended high yielding varieties, treat the seed with Captan or Thiram (3 g /kg seed)
- Grow wilt no need of, stunt and dry root rot resistant varieties
- Follow IPM for the management of *Heliothis*
- Intercultivate twice at 20 and 30 DAS
- Spray Fluchloralin at 2.5 I/ha as pre-sowing incorporated or spray Pendimethalin at 3.3 to 5 I/ha immediately after the sowing or next day.

Soybean (Glycine max)

- Use recommended high yielding varieties
- Treat the seed with *Rhizobiiun japonicum*
- Grow soybean as an intercrop in cotton. This not only increases the area under soybean but also helps in the buildup of predators of Helicoverpa
- Cultivate soybean in rabi under ID conditions.

Groundnut (Arachis hypogaea)

- Intercrop with redgram/castor/bajra/jowar in 7:1 ratio
- Deep summer ploughing
- Sow boarder crop (4 rows) of jowar or bajra
- Adopt quality seed of HYV
- Adopt recommended improved variety, plant population (33/m^2) and plant protection
- Seed treatment and use small seed without shrivelling
- Do not grow sunflower and marigold which are highly susceptible to tobacco streak virus that causes peanut stem necrosis.

- Use graded seed, follow seed treatment and adopt recommended seed rate.

- Use gypsum (dose 500 kg/ha in the next ploughing or at 45 DAS) and SSP to provide calcium and sulphur

- Avoid delayed groundnut sowings beyond second fortnight of July which will result in increased pest and disease problems and reduced yields

- Practice crop rotation and inter cropping

- Use mechanization for sowing, inter cultivation, harvesting and strippling to reduce cost of cultivation.

Castor (Recinus cammunis)

- Use recommended variety/hybrid

- Follow seed treatment (3 gm/kg of seed with Thiram or captan)

- Apply recommended fertilizers, N in 2-3 splits

- Control RHC semilooper, spodoptera through IPM

- Intercrop with redgram and cowpea

- Remove and destroy botrytis affected spikes and spray 0.1% carbendazim solution on the remaining spikes. Apply 20 kg N and 10 kg K_2O/ha if there is moisture in soil

- Popularize rabi castor in non-traditional areas.

Sesamum (Sesamum indicum)

- Use quality seed of recommended varieties, seed treatment is essential

- Sow timely in rows, 2 kg seed mixed with 6 kg sand/ac with seed drill, after 2-4 ploughngs and leveling with 2 harrowings, adopting 30 x 15 cm spacing

- Complete thinning and weeding by 15-20 days after sowing

- Apply fertilizers in split doses. Basal application of 16 : 24 : 8 kg NPK/ac and 20-25 kg urea at 30 DAS

- Go for timely control of leaf folder and gall fly

- Spray wet table sulphur @ 2.0 g/litre of water to control powdery mildew
- Destroy phyllody affected plants.

Sunflower (Helianthus annus)

- Use quality seed of recommended varieties/hybrids, seed treatment with 2-3 grams of thiram/captan per kg of seed
- Apply recommended fertilizers; N in 2-3 splits
- Don't use any insecticide during flower opening
- Protect crop from Helicoverpa, Spodoptera and parrots. To control spodoptera spray neem oil in early stage @ 5 ml/Lt or monocrotophos 2.0 ml/Lt.

Cotton (Gossypium sps)

- Resort to summer ploughing. Avoid mono cropping of cotton
- Use delinted and treated seed of recommended varieties /hybrids
- Go for intercropping with soybean/greengram/cowpea to reduce pest load/improve soil fertility and net returns
- Adopt stem application of systemic insecticides at early stage to encourage defender population
- Collect and destroy eggs of helicoverpa and spodoptera and grown up caterpillars
- Follow IPM schedule
- Remove cotton stubbles soon after final picking
- Do not grow cotton in light soils and chalka and dubba soils as a rainfed or even as an irrigated crop
- Seed cotton from fully opened bolls should be collected during cooler times of the day
- Seed cotton damaged by bollworms should be picked separately
- Seed cotton should be graded and stored in heaps or in gunny bags in dry and well ventilated godowns

- Watering the seed cotton before weighment should be avoided
- Mixing of seed cotton of different varieties should be avoided
- Proper packing should be done to protect from contamination, dampness and fiber quality.

Sugarcane (Saccharum officianarum)

- Use short crop for seed material. Soaking setts in 10% lime solution for 60 minutes is beneficial for better seedling establishment and growth under limited irrigation sources
- Give hot water treatment (52°C for 30 minutes) followed by chemical treatment of seed material (0.1% malathion + 0.05% carbendazim)
- Gap filling has to be done within 2 weeks of ratooning
- Apply P as basal and N and K fertilizers in two splits in recommended doses
- Irrigate at weekly intervals at formative phase during summer
- Ratoon crop management
    - o Take up stubble shaving, gap filling, interculture, higher dose of nitrogen, foliar spraying of $FeSO_4$
    - o Apply potassic fertilizers to rainfed crop (120 kg ./ha) at planting and at 90 DAP
    - o Adopt trash mulching @ 3 t/ha at 3 rd day after planting, or immediately after rationing to conserve moisture
    - o Irrigate once at 10 DAP (life saving irrigation) and second at 30 days thereafter under rainfed conditions.

Early Shoot Borer

- Adopt trash mulching 3 t/ha at 3rd day after planting
- Apply phorate 10 G @ 15 kg/ha at the time of planting
- Irrigate at closer intervals during summer
- Spray chloropyriphos 20Ec @ 2.5 ml/Lt of water at 4th, 6th, 9th weeks after planting.

Scale insect

- Dipping setts in 0.1% malathion or dimethoate (0.05%) before planting
- Detrash keeping eight green leaves at the top followed by Malthion (0.1%), Chlorpyrifos (0.05%) or Dimethoate (0.05%) in first week of July, August and September.

Red Rot

- Use red rot resistant varieties.

Smut

- Give hot water treatment of seed material
- Avoid second ratoon, from the smut affected plant crop.

Chillies (Capsicum fruitisense)

- Grow only virus resistant varieties
- Treat the seed with Mancozeb or Captan or Bavistin @ 3 g/kg of seed
- Drench with Bordeaux mixture (1%) or Copper oxychloride (3 g/1) on 13th and 20th day of sowing to prevent damping off diseases
- Follow Integrated Nutrient Management
- Avoid indiscriminate use of pesticides to save the crop from phytotoxicity and residues on the produce
- Avoid repeated use of synthetic pyrethroids as it leads to secondary infestation of sucking pests
- Spray 1% urea 3-4 times at fortnightly interval under moisture stress / condition.

7.11 Policy Options for Drought Governance

National Academy of Agricultural Sciences (2013) has identified policy options and actions for climate change and governance of drylands. They include;

- Standardise methodologies for vulnerability assessment and climate smart agriculture; enhance the density of weather observatories; establish rain gauges at block/village level;

and enable access to and efficient management of weather related information by modern tools like remote sensing and GIS.

- Institutionalize a mechanism to collect and collate micro-level information continuously (climate, crops, socio-economic, natural resources etc.) and establish credible database as well as efficiently disseminate it which can be used as an input for macro-level policies.

- Increase Investment in research and development of locally-adaptable crops, management and models for analysing the impact of climate change and mitigation strategies, particularly in dryland regions in view of their greater vulnerability.

- Develop and diffuse drought, heat and submergence tolerant crop cultivars. Change planting dates to avoid terminal heat stress at maturity stage.

- Improve techniques for water (rainwater harvesting, micro-irrigation), nutrient (SSNM) and energy (reduced tillage) conservation to adapt the crops to climatic stresses and also contribute to mitigation.

- Encourage adoption of location-specific conservation techniques (cover cropping, *in situ* moisture conservation, rainwater harvesting, groundwater recharge, locally adapted cropping systems etc.) for water efficient agriculture and demonstration of such technologies on farmers' fields.

- Blend farmers' traditional/indigenous knowledge on natural resource management and climate coping strategies with advanced technological interventions e.g., varieties, crop management, resource conservation technologies, rainwater harvesting and storage.

- Manage climate risks through weather-based agro-advisories, and affordable weather insurance products. The Government of India should move from reactive (relief payments) to practice (promoting insurance) approach in dealing with climate variability.

- Integrate climate change initiatives such as INCCA, NAPA, NMSA, NICRA, NDMA etc., with national agricultural policies/programmes of food security, disaster management, natural resource conservation, livelihood enhancement etc., to enable farmers take advantage of new technologies and tools.

- Encourage diversification of rural income through off-farm and non-farm livelihoods; build significant stake for climate change adaptation interventions in the newly-launched national livelihood mission.

- Ensure access to government support/relief programs (food security, agricultural and enterprise subsidies, rural finances, poverty reduction programs, technology adoption support etc.) on part of the vulnerable sectionsof the society.

- Create favourable environment to attract public and private funds for investment in Climate Resilient Agriculture (CRA) or Climate Smart Agriculture (CSA). Much of the corporate social responsibilities funds from leading private sector players in the country should be channelized to promote CRA CSA in vulnerable regions.

- Channelize investment in human resource development and capacity building through awareness and training among officials, extension workers and farmers, incentives to farmers for adoption of natural resource conservation practices and support to improve the existing indigenous technologies that are eco-friendly and sustainable in long-run.

- Revise and restructure university curricula and institutionalise teaching and research facilities in concerned educational institutions to develop graduates and postgraduates suitably trained in climate change management.

- Encourage role of the Non-Governmental Organizations, Civil Societies, Public and Philanthropic organizations for enhancing adaptation preparedness among the local community.

- Increase concessional credit to small and marginal farmers for adoption of climate resilient agricultural practices. Evolve a new verification system at field level to sensitise farmers through credit and input subsidies.

- Establish efficient co-operatives, farmer producer organizations (FPOs), Commoditi Interest Groups (CIGs), community organizations and associations/groups to tackle critical needs of farmers like resource mobilization, custom hiring, marketing their outputs and efficient natural resource management.

- Evolve national policy on disaster management in agriculture when major events like droughts, cyclones and floods devastate agriculture, horticulture and livestock.

- Implement the recommendation of the National Commission on Farmers (NCF) for establishment of a National Agricultural Risk Management Fund and its need-based timely distribution.

- Identify, critically document and share most successful stories in climate change management and organise pilot demonstrations of most successful modules, and in a partnership mode scale-up and scale-out the selected ones to establish climate smart villages throughout the country.

- Introduce soil and water conservation practices on a large scale for achieving long-term sustainability of the systems, including revitalization and management of Common Property Resources. Farmers must be given incentives to adapt CRA practices.

- Build on priority infrastructure like markets and information gateways and create opportunities in the non-farm sector in vulnerable regions to help the farmers to diversify their incomes.

- converge the many programmes/projects on climate resilient agriculture funded by a multitude of ministries and government agencies. Obviously, there is a need to synchronize and implement the developmental programs with built-in monitoring and a participatory bottoms-up feedback loop in the system for meaningful climate resilience *per se* in agricultural sector in a mission mode.

Conclusion

The dryland agriculture depending on rains will have to play a crucial role in sustainable food security and enhanced livelihoods of small and marginal framers. The resilience in the face of adverse effects of climate change on small holder agriculture needs to be developed that could bring in sustainability of farming systems. The challenge before us, therefore, is to prepare for drought planning and mitigation and continue to develop management practices that increase climate adaptability to varying climate scenario and built resilience of the farming communities and farming systems. Resilience to predicted climate change will depend on increasing agricultural productivity with available water resources, refining technologies and timely deployment of affordable strategies to accomplish potential yield levels of drylands land and water productivity.

References

Channabasavanna AS., Biradar, D.P., Prabhudev, K.N., and Hegde, M. (2009). Development of Profitable Integrated Farming System Model for Small and Medium Farmers of Tungabhadra Project Area of Karnataka. Karnataka Journal of Agricultural Sciences, 22(1): 25-27.

Davies, J., Poulsen., L, Schulte-Herbrüggen, B., Mackinnon, K., Crawhall, N., Henwood, W.D., Dudley, N., Smith, J. and Gudka, M. (2012) Conserving Dryland Biodiversity. IUCN –Drylands Initiative, Nairobi.

Dawson, T.P., Jackson, S.T., House, J.I., Prentice, I.C. and Mace, G.M. (2011). Beyond predictions: biodiversity conservation in a changing climate. Science 332: 53–58.

FAO (2008) Conservation Agriculture. 2 008-07-08 http://www.fao.org /ag/ca/ index.html.

Hobbs, P.R. (2007). Conservation Agriculture: What is it and Why is it Important for Future Sustainable Food Production? J. Agric. Sci. 145, 127–137p.

Hobbs, P.R., Sayre, K. and Gupta, R. (2008). The role of conservation agriculture in sustainable agriculture. Philos. Trans. Roy. Soc. B 363, 543–555p.

IPCC (2007). Impacts, Adaptation and Vulnerabilty. The intergovernmental Panel on Climate Change, Cambridge University Press, U.K.

Jayanthi, C., Rangasamy, A. and Chinnusamy, C. (2000). Water Budgeting for Components in Lowland Integrated Farming Systems. Agricultural Journal, 87: 411- 414p.

Joshi, P.K. and Pramod Aggarwal. (2014). Agriculture must get climate smart. Financial Express 19, May 2014.

Jung, C. and Müller, A. (2009). Flowering Time Control and Applications in Plant Breeding. Trends Plant Sci. 14, 563–573p.

Kavi Kumar, K.S. (2011). Climate Sensitivity of Indian Agriculture: Do Spatial Effects Matter? Cambridge Journal of Regions, Economy and Society, 4(2): 221-235p.

Nkonya, E., Winslow, M., Reed, M. S., Mortimore, M., & Mirzabaev, A. (2011). Monitoring and Assessing the Influence of Social, Economic and Policy Factors on Sustainable Land Management in Drylands. *Land Degradation & Development*, 22(2), 240-247p.

Parida, A. (2012). Role of Agriculture in Alleviating Poverty and Malnutrition – Presidential Address, Agriculture and Forestry Section. Indian Science Congress Association. *Proc. Agril Forestry Section*, ISCA, Kolkata, 327p.

Pathak, P., Sahrawa,t K.L., Wani, S.P., Sacan, R.C. and Sudi, R. (2009). Opportunities for Water Harvesting and Supplemental Irrigation for Improving Rain-Fed Agriculture in Semi-Arid Areas. In *Rain-fed agriculture: Unlocking the Potential.* (S.P. Wani, John Rockstrom and Theib Oweis Eds.).Comprehensive Assessment of Water Management in Agriculture Series, CAB International, Wallingford, UK, 197-221p.

Peterson, T.C., Hoerling, M.P., Stott, P. A. and Herring, S. (2013). Explaining Extreme Events of 2012 From A Climate Perspective. Bull. Am. Meteorol. Soc. 94, S1–S74.

Saseendran, R.M., Smith, I.M. and Matson, P.A. (2000). Ecological and Evolutionary Responses to Climate Change. Science 284: 1943-1947p.

Sinha, S.K. and Swaminathan, M.S. (1991). Deforestation, Climate Change and Sustainable Nutrition Security: A Case Study of India. Climate Change 19: 201-209p.

Singh, S.N., Saxena, K.K., Singh, .KP., Kumar, H. and Kadian, V.S. (1997). Consistency in income and employment generation in various farming systems. Annals of Agril. Res., 18(3): 340-43p.

Sivakumar, M.V.K. (2012). High-level Meeting on National Drought Policy. CSA News, December. American Society of Agronomy, Madison, Wisconsin, USA.

Thomas, R.J., Eddy de Pauw, Manzoor Qadir., Ahmed Amri., Mustapha Pala., Amor Yahyaoui., Mustapha El-Bouhssini., Michael Baum., Luis Iñiguez and Kamel Shideed (2007). Increasing the Resilience of Dryland Agro-ecosystems to Climate Change. SAT eJournal 4: 1-17p.

Venkateswarlu, B and Shanker, A.K. (2009). Climate change and agriculture: adaptation and mitigation strategies. Indian J Agron 54: 226–230p.

Wani, S.P., Sreedevi, T.K., Rockstrom, J. and Ramakrishna, Y.S. (2009). Rain-fed Agriculture – Past Trend and Future Prospects. In *Rain-fed agriculture: Unlocking the Potential. Comprehensive Assessment of Water Management in Agriculture Series* (S.P. Wani., J. Rockstrom and T. Oweis, Eds.). CAB International, Wallingford, UK. 1-35p.

Wilhite, D.A., Mannava V.K. Sivakumar. and Roger Pulwarty. (2014). Managing Drought Risk in a Changing Climate: The Role of National Drought Policy.

Wilhite, D.A., Pulwarty, R.S. (2005). Drought and Water Crises: Lessons Learned and the Road Ahead. In: Wilhite, D.A. (Ed.), Drought and Water Crises: Science, Technology and Management Issues. CRC Press, Boca Raton, Florida, 389–398p (Chapter 15).

8

Climate Smart Agriculture for Building Resilience and Improving Livelihoods in Rainfed Areas

Suhas P Wani

ICRISAT Development Centre,
International Crops Research Institute for the
Semi-Arid Tropics (ICRISAT), Hyderabad.

8.1 Introduction

The biggest challenge, human kind facing in the 21st century is to cope up with the impacts of the climate change which is affecting the sustainable development globally. The IPCC (Inter Governmental Panel on Climate Change) has clearly established that climate change is reality now and going to affect food security and sustainability in different regions (IPCC, 2013). Our generation is the first generation to assess the impacts of the climate change and also is the last generation which can do interventions to minimize/reverse the climate change on the globe. The dryland agriculture which is globally 80% and contributes 60% of the food is the most vulnerable systems for the impacts of the climate change. Dryland areas are also the hot-spots of poverty in developing countries of Asia, Africa and Latin America and are more vulnerable to the adverse impacts of climate change. In order to sustain the growth globally as well as the livelihoods and achieve the goals of food and nutritional security, there is an urgent need particularly in thickly populated countries like India and China. In India, 58% of arable land is dryland agriculture and will be severely affected as 1/3rd of the developing world will be facing physical scarcity of water by 2030 (Rockstorm et al., 2007). There is an urgent need to develop climate smart agriculture by adopting appropriate adaptation and mitigation strategies to build the resilience of the systems as well as livelihoods for small farmholders in the country.

International Crops Research Institute for the Semi-Arid Tropics (ICRISAT) is a global research and development organization addressing the issues of semi-arid tropics. Our Vision is "A prosperous, food-secure and resilient dryland tropics" and our Mission is "To reduce poverty, hunger, malnutrition and environmental degradation in the dryland tropics". ICRISAT in partnership with National Agricultural Research Systems, Development Agencies, Civil societies and farmers organizations is undertaking research for development to address the issues of climate change, water scarcity, land degradation and food insecurity for ever growing population which has already reached 1.2 billion in the country. The climate change scenario analysis for South Asia has demonstrated increased temperature during 2010-2039 by 0.54° to 1.17° and 1.71° to 3.16° during 2040-2069 with insignificant changes in precipitation. Although, total quantity of precipitation may not be affected but the intensity and distribution of the rains will be affected severely as already experienced during this decade itself. There are planetary boundaries for safe operation humanity and out of nine parameters already we have crossed the threshold limits for three of them viz., biodiversity loss, nitrogen cycle and climate changewhich have reached beyond its permissible threshold at planetary scale. Recent analysis of changes in space of agro-eco regions in the country using weather data sets from 1971-1990 and 1991-2004 have integrated increased semi-arid areas by 8.45 million ha with anoverall increase of 3.45 million ha addition to the semi-arid tropical areas, mainly in the states of Madhya Pradesh, Bihar, Uttar Pradesh, Karnataka and Punjab. At micro level looking at different watersheds, clearly impacts of climate change in terms of reduced number of rainy days, increased frequency of occurrence of dry spells and high intensify rains as well as reduced length of growing period have been observed.

8.2 Adaptation and Mitigation Strategies

To cope with the climate smart agriculture is based on the principle of integrated adaptation strategies along with mitigation measures for achieving the food security and sustainable development. Smart agriculture calls for a holistic approach and its success depends on building the partnership between knowledge-generating institutions (research organizations) with knowledge-transforming institutes (extension agencies) with development programs of investors and governments through scaling-up of the proven approaches. Some of the adaptation measures of climate smart agriculture include improved rainwater conservation and management along with redeployment and retargeting of the available germplasm of climate ready crops and application of existing knowledge on diversification of crops/systems for improving the livelihoods. A catchment management approach through integrated watershed management with community participation has proven its ability to cope with the impacts of the climate change by bridging the yield gaps between current farmers' productivity and achievable potential in different eco regions (Wani et al., 2012). We refer to this initiatives as "Hypothesis of Hope" for developing adaptation strategies to cope with the impacts of the climate change and in due course by developing climate smart crop cultivars we can definitely minimize the impacts of the climate change.

We need to build the resilience and reduce the vulnerability for which there is a need to understand the concept of resilience and vulnerability. Through building the capacity of stakeholders as well as sensitizing the farmers and the policy makers using new science tools suitably. The case studies of selected watershedexperiences in the country as well as from other countries like China, Thailand and Vietnam will be discussed as examples to show how the resilience has been built for keeping with the impacts of the climate change through integrated watershed management experiences of scaling-up of technologies for achieving the impacts of large population through outcomes will be analyzed of the mission program like Bhoochetana in Karnataka and Andhra Pradesh as well as an integrated mission approach for improving livelihoods through climate smart agriculture through programs like "Bhoo Samrudhi" in Karnataka and "Rythukosam" in Andhra Pradesh will be discussed. Drivers of success for scaling-up programs will be shared with the participants.

References

IPCC, (2013). Climate Change 2013: The Physical Science Basis. Contribution of Working Group I to the Fifth Assessment Report of the Intergovernmental Panel on Climate Change [Stocker, T.F., D. Qin, G.-K. Plattner, M. Tignor, S.K. Allen, J. Boschung, A. Nauels, Y. Xia, V. Bex and P.M. Midgley(eds.)]. Cambridge University Press, Cambridge, United Kingdom and New York, NY, USA, 1535p.

Rockström, J., Nuhu Hatibu, Theib Oweis. and Wani, S.P. (2007). Managing Water in Rain-fed Agriculture. In: Molden, D., (ed.) Water for Food, Water for Life: A Comprehensive Assessment of Water Management in Agriculture, Earthscan, London, UK and International Water Management Institute (IWMI), Colombo, Srilanka, 315-348p.

Wani, S.P., Garg, K.K., Singh, A.K. and Rockstrom, J. (2012). Sustainable Management of Scarce Water Resources in Tropical Rainfed Agriculture. In: Soil Water and Agronomic Productivity. Advances in Soil Science, CRC Press., 347-408 p.

9

"Climate Resilient Agricultural Practices – NICRA Experiences"

Y.G. Prasad
*Principal Scientist & Coordinator,
NICRA Technology Demonstration Component
ICAR-CRIDA, Hyderabad*

9.1 Introduction

ICAR launched a nation-wide project "National Initiative on Climate Resilient Agriculture (NICRA)" in 2011 to carry out basic, applied and strategic research to address climate change and climate variability and work towards imparting resilience to Indian agriculture. As part of this initiative, the Technology Demonstration Component (TDC) addresses the issues of enabling farmers through demonstration of available location-specific climate resilient practices and technologies. The demonstrations related to natural resource management, crop production, livestock and fisheries offer an integrated package of proven, location-specific technologies for adaptation of the production systems to climate variability. TDC is under implementation in a farmer participatory mode in 130 vulnerable districts of the country through 121 Krishi Vigyan Kendras (KVKs) and 7 TOT divisions of ICAR research institutes.

The demonstrations are categorised broadly under four different modules viz., natural resource management, crop production, livestock and fisheries and institutional interventions. To facilitate their effective implementation, various institutional mechanisms were also established at the village level.

9.2   Natural Resources

This module consists of interventions related to water harvesting and recycling for supplemental irrigation, improved drainage in flood prone areas, conservation tillage where appropriate, artificial ground water recharge, and efficient water management methods. Over 800 Rain Water Harvesting (RWH) structures were constructed, renovated, or repaired in the project. An additional rainwater storage capacity of two lakh cu.m was created through farm ponds alone.

9.3   Crop Production

This module consists of introducing adaptation technologies and practices to impart resilience to crop production. The interventions include demonstration of short duration drought or flood tolerant crop cultivars, contingency crop planning, soil test based nutrient application, water saving methods (direct seeding), advancement of planting dates of rabi crops in areas with terminal heat stress, in situ moisture conservation and crop diversification.

9.4   Institutional Interventions

This module consists of institutional interventions by either strengthening the existing ones or initiating new ones especially an innovative farm machinery custom hiring centre (CHC). Timeliness of agricultural operations is crucial to cope with climate variability, especially with sowing and intercultural operations.

Land shaping for rainwater harvesting, utilization & integration of farm enterprises: In order to overcome submergence during *kharif,* salinity problem in *rabi* and augment availability of irrigation water during the *rabi*-summer season, an engineering solution was promoted by the KVK, scientist. The excavated soil was spread over the adjacent field area (80%) so as to raise the field level up to 1 to 1.5 feet and also to raise the field border embankments of land. Fish and duck rearing were introduced in the harvested freshwater in the pond. Thus, this technology of land shaping offered a model for harvesting rainwater in *kharif,* vegetable cultivation during *rabi* and fresh water fish culture in the ponds.

Establishment of community paddy nursery: Establishing a staggered community nursery was explored as a local adaptation strategy at the village level to combat the problem experienced by farmers during deficit rainfall seasons in lowlands. The technique involves raising a staggered community nursery under assured irrigation in the village at an interval of 2 weeks. In the anticipation of a two weeks delay in monsoon the first nursery is taken up as a contingency measure by 15 June with the long duration variety (>140 days) in order to transplant 3-4 weeks old seedlings by first fortnight of July. If the monsoon delay extends by 4 weeks, the second nursery is raised with medium duration varieties (125-135 days) by 1st July to supply 3-4 weeks old seedlings for transplanting in the 3rd or 4th week of July. In case of anticipation of further delay or deficit rainfall conditions, the 3rd nursery is raised by mid July with short duration varieties (<110 days) to take up transplanting of 3-4 week seedlings in the first fortnight of August.

Direct seeded rice in un-puddled field to cope with water shortages: Direct seeding of drought tolerant varieties of rice in dry soil is done in June with pre-emergence herbicide application (pendimethalin 1 kg/ha) under sufficient soil moisture conditions followed by a post-emergence herbicide application (bispyribac sodium 25g/ha) at 25-35 days after sowing or hand weeding at 35-45 days after sowing to effectively manage weed problem. Direct seeding of rice is done with a zero till drill. The quantity of seed required is 20- 25 kg/ha compared to transplanted paddy which requires 60-80 kg/ha.

Drum seeding of rice for water saving and timeliness in planting: Drum seeding technique involves direct seeding of pre-germinated paddy seeds in drums made up of fiber material to dispense seeds evenly in lines spaced at 20 cm apart in puddled and levelled fields. About 35 to 40 kg paddy seed/ha is soaked overnight in water and allowed to sprout. Care should be taken not to delay sowing as seeds with long shoot growth are not suitable for drum seeding. The sprouted seed is air-dried in shade briefly (<30 minutes) prior to sowing for easy dispensing through the holes in the drum seeder.

Drought tolerant paddy cultivars to tackle deficit rainfall situations: Short duration and drought tolerant varieties that can withstand up to 2 weeks of exposure to dry spells in rainfed areas were demonstrated in NICRA villages. Drought tolerant cultivars demonstrated in farmers fields include: 'Sahbhagi dhan' (105-110 days duration in plain areas and 110-115 days in uplands, highly resistant to leaf blast and moderately resistant to brown spot and sheath blight. 'Naveen' (115-120 days duration and 'Anjali' (90 days). Other early maturing varieties that have potential in the eastern states include: 'Birsa Vikas Dhan 109' (85 days duration), and 'Abhishek' (120 days duration).

Short duration finger millet varieties for delayed monsoon/deficit rainfall districts in south interior Karnataka: When the delay in monsoon is about 4 weeks, medium duration varieties like GPU-28 (110 days duration) performed better while in case of further delay, short duration varieties like ML-365 (105 days) and GPU-48 (100 days) performed better. These short duration varieties of finger millet are also tolerant/ resistant to blast disease and can be sown till August under rainfed conditions in medium to deep red soils.

Crop diversification for livelihood security and resilience to climate variability: Pigeonpea, cotton, sunflower and sorghum are the main crops cultivated in NICRA village in Kurnool district. These crops are affected due to late onset of monsoon followed by dry spell at critical crop growth stages. Intercropping of Setaria (foxtail millet, SIA-3085 variety) with pigeonpea (5:1 ratio) sown in July showed that the intercropping system was more profitable with highest benefit cost ratio in all the 3 years despite prolonged dry spell of up to 25 days. In other semi-arid agro-ecosystem, intercropping of soybean + pigeonpea (4:2), pearlmillet + pigeonpea (3:3), pigeonpea + green gram (1:2) and cotton + green gram (1:1) performed significantly better than their sole crops in case of deficit rainfall conditions.

Flood tolerant varieties impart resilience to farmers in flood-prone areas: Rice varieties   Swarna-sub1, MTU-1010, MTU-1001 and MTU-1140 are high yielding with good grain quality apart from possessing submergence tolerance and perform better under flood situation. Demonstration of these varieties in flood-prone areas showed that Swarna-sub1, a variety developed by IRRI and CRRI, Cuttack and released in 2009, could tolerate submergence up to two weeks and could perform significantly better compared to other improved and local cultivars. MTU-1010 is a short duration, dwarf variety resistant to lodging and can withstand moderate wind velocity. This attribute of lodging resistance saves from not only loss in grain but also straw yield which is the main source of dry fodder. MTU-1140 is also a promising, non- lodging variety comparable in grain quality to BPT-5204, which performed well in flood-prone villages in Srikakulam district of Andhra Pradesh.

Community tanks/ponds as a means of augmentation and management of village level water resources: One of the first works accomplished in NICRA villages was to identify these structures to carry out desiltation with farmer's participation. The rich silt deposited in these structures was used by farmers for spreading in the fields, wherever necessary, to improve the water holding capacity of soils. This intervention helped in increasing the surface water resource availability, increased the ground water recharge observed through water table measurements in wells located nearer to the tanks.

Individual farm pond for improving livelihood of small farmer: One way to cope with climate vulnerability is to collect rainwater in harvesting structures to increase the irrigated areas as well as crop productivity. Farm ponds have been considered as one of the key interventions in NICRA villages in Karnataka, Telangana, Andhra Pradesh and Maharashtra and have been widely adopted in the villages. Various cropping system modules were worked out by using harvested water. Majority of farmers opted to cultivate vegetables with harvested water in a ratio of 1:10 (command to catchment area) with sustained profits.

Low Cost water harvesting structure-Jalkund: The shortage of water can be minimized by rainwater harvesting and its judicious use for crop production. Direct rainfall collection through water catch ponds/pits (Jalkund) or storage by diverting runoff can be highly beneficial to farmers for providing irrigation to crops during moisture scarcity conditions during dry seasons. Rainwater can be stored directly in *Jalkunds* during the rainy season which can be utilized to provide protective irrigation to the crops for successful cultivation.

Storing excess runoff in streams through check dam: *Ex-situ* storage of water in seasonal streams at suitable sites is an important strategy to conserve excess runoff water in different rainfall zones. Often, by virtue of the location with reference to nearby hilly areas, the village may receive copious amounts of surface runoff from surrounding areas. This excess runoff could be harvested in streams either for direct use or for improving the ground water availability. In high rainfall areas, though runoff availability is high, often it gets lost due to non-availability of storage structures. In these regions, on-stream storage structures could be built on first order streams to make water available for direct use during long dry spells by farmers. In majority of NICRA villages, check dams (new/desilting of existing ones) were major interventions in drought prone districts in different rainfall zones.

Sand bag check dam: Rainwater harvesting and recycling was demonstrated by construction of temporary check dams. Temporary check dams were constructed by using low cost gunny/polythene bags. These bags were filled with sand from stream beds. The sand filled bags were placed one above the other in two or more rows. The gap between two rows was filled with clay to check water leakage. An outlet was provided to each dam to allow excess water to flow downstream. These check dam helps in ground water recharge and rising of water table in the area. Harvested water in the temporary check dams is used for life saving irrigation in *rabi* and summer crops.

Recharge of wells: Recharging of tube wells was taken up as a major intervention. The technique involved diverting runoff to a pit dug around the tube well after trapping the silt. About 8 to 10 cement rings were descended into the dugout around the tube well. Harvested rainwater is collected in a cement tank and allowed for the silt to settle down and then conveyed to the dugout using a PVC pipe. Material required for recharging included cement rings (8-10 numbers, 4.5 ft radius, 2 ft height), bricks (500), cement (1 bag), pipe (10 ft long) and one perforated pipe (6ft long). Average cost incurred was Rs.10,000 out of which 25 per cent of total cost and labor was shared by the farmer.

Integrated farming systems: Several integrated farming system modules with a combination of small enterprises such as crop, livestock, poultry, piggery, fish and duck rearing were demonstrated to farmers in NICRA villages in the eastern, northern, north eastern and southern states where monocropping is mostly practiced due to climatic conditions.

Captive rearing of fish seed - a livelihood opportunity in flood-prone areas: Captive rearing of fish seed i.e., rearing of early stages (spawn to fry and fry to fingerling stages) through appropriate feed and health management in nursery pond was demonstrated in Sirusuwada village, Kothur mandal of Srikakulam district, Andhra Pradesh. Required training for captive rearing was provided to the fisher folk by the KVK. The practice brought about self-reliance along with reducing costs and increasing returns by adequate stocks in main tanks.

Management practices to tackle cold stress in backyard poultry: A brooder house was made using locally available materials such as bamboo and wood and to maintain optimum night temperature in the shelter with the help of light bulbs during cold stress period. Breed improvement was made with the introduction of Gramapriya and Vanaraja chicks in NICRA villages. The movement of chicks was restricted nearer the heat source in brooder house with the help of chick guards made with card board. The chicks were fed with ground maize initially, later fed with vegetable wastes and other kinds of locally available grains like maize and rice bran besides the feed material available from free range scavenging as they grew up. Wholesome fresh water was made available at all times in the watering and feeding trough made of bamboo. Chicks were also supplemented with multivitamins @ 1ml/lt of water.

Shelter management: A semi-intensive system of rearing of goats in a slatted floor with proper roof can provide shelter to the animals to tackle heat stress during summer and rain storms during monsoon. Locally available wooden planks were used for making slatted floor. A gap of one inch was maintained between each slat so that the urine and fecal material could be collected from the bottom of the floor. The roof of the shelter was made with bamboo and covered with either thatched material or coarse cereal crop residues. The improved shelter reduced kid mortality and increased profitability.

Improved planting methods for enhancing water use efficiency and crop productivity: There is a need for *in-situ* soil and water conservation and proper drainage technology in deep black soils. Broad bed and furrow (BBF) system involves preparation of a broad bed of 90 cm, furrow of 45 cm and sowing of crop at a row spacing of 30 cm. The cost of BBF implement is Rs. 45,000. The BBF technology has many advantages including *in-situ* conservation of rainwater in furrows, better drainage of excess water and proper aeration in the seedbed and root zone. More than 200 farmers in Sanora and Barodi village in Datia district of MP adopted the technology. Similarly, furrow irrigated raised bed (FIRB) planting was promoted for cultivation of different crops in several states viz., Uttar Pradesh, West Bengal, Punjab, Maharashtra, Karnataka, Rajasthan and Tamilnadu. Ridge and furrow method of vegetable cultivation was promoted in Gumla district and in cotton at Amravati and Aurangabad, Maharashtra.

Zero till planting of rabi crops: Demonstrations of zero till drill sown wheat in farmers' fields were undertaken in several NICRA villages in Punjab, Haryana and Uttar Pradesh. The zero till drill not only saved tillage costs and energy but also eliminated the need for seedbed preparation. Zero till drilled wheat yields were on par with conventionally sown wheat. The machine operated with a 35 hp tractor to cover sowing of wheat in 4-5 ha/day.

In situ incorporation of biomass and crop residues for improving soil health: In order to encourage farmers to change this practice, rotavator machine was introduced in several NICRA villages. The harvested crop stalks/stubbles are chopped into small pieces and incorporated *in-situ* into the soil with varying efficiencies depending upon the left over residue. The cost of implement is Rs. 1.0 to Rs. 1.2 lakhs and field capacity of the rotavator is 5-6 ha/day. Rotavator helps in obtaining of early seedbed preparation soon after harvesting of *kharif* crops for sowing of *rabi* crops. This not only requires low energy in tillage operation but also mixes and incorporates the stubbles of previous crop thoroughly in the soil. This improves the soil physical properties and hence, results in increased crop yield. Incorporation of green manuring crops such as daincha, moong and cowpea in wet conditions can be taken up to improve soil health.

Village level seed banks to combat seed shortages: Participatory village level seed production of short duration, drought and flood tolerant varieties was demonstrated in several NICRA villages with the support of KVKs in rice, soybean, groundnut, greengram, finger millet, foxtail millet and pigeonpea. Breeder seed/ foundation seed was sourced from research farms for multiplication in farmers fields and the quality seed so produced was mostly used in the village and nearby villages. Farmer to farmer sale as truthful seed was the means of spread.

Fodder cultivars to tackle fodder scarcity: Short and medium duration fodder cultivars of several crops that can withstand up to 2-3 weeks of exposure to drought in rainfed areas were demonstrated in NICRA villages. These include: sorghum (Pusa Chari Hybrid-106 (HC-106), CSH 14, CSH 23 (SPH-1290), CSV 17); Bajra (CO 8, TNSC 1, APFB 2, Avika Bajra Chari (AVKB 19); Maize (African tall, APFM 8). These cultivars can be sown immediately after the rains under rainfed conditions in arable lands during *kharif* season and are ready for cutting by 50-60 days. Cultivars of *rabi* crops like Berseem (Wardan, UPB 110) and Lucerne (CO 1, LLC 3, RL 88) were demonstrated in NICRA villages as second crop with the available moisture during winter. Perennial fodders like APBN-1, CO-3 and CO-4 were also demonstrated under limited irrigated conditions.

Conclusion

Climate resilient agricultural practices need to be adopted widely by farmers for risk management and enhancing adaptive capacity. This needs awareness creation and exposure through participatory technology demonstrations. Various barriers for adoption such as lack of adequate support for timely availability and access to resources, machinery and inputs need to be effectively addressed. Some of these practices can be scaled-up and scaled-out with support from various ongoing government schemes such as the National Mission on Sustainable Agriculture and Green Climate Fund.

References

Prasad YG, Venkateswarlu B, Ravindra Chary G, Srinivasa Rao Ch, Rao KV, Ramana DBV, Subba Reddy G, Rao VUM and Singh A K. (2012). Contingency Crop Planning for 100 Districts in Peninsular India. CRIDA, Hyderabad. (ISBN: 978-93-80883-13-07). 302p.

Venkateswarlu B, Singh AK, Prasad YG, Ravindra Chary G, Srinivasa Rao Ch, Rao KV, Ramana DBV and Rao VUM. (2011). District Level Contingency Plans for Weather Aberrations in India. CRIDA, Hyderabad (978-93-80883-07-6). 136p.

Srinivasarao Ch, Ravindra Chary G, Mishra PK, Nagarjuna Kumar R, Maruthi Sankar GR, Venkateswarlu B and Sikka AK. (2013). Real Time Contingency Planning: Initial Experiences from AICRPDA. All India Coordinated Research Project for Dryland Agriculture (AICRPDA). Central Research Institute for Dryland Agriculture (CRIDA), ICAR, Hyderabad-500059, India. 63p.

Ch. Srinivasa Rao, B. Venkateswarlu, A.K. Sikka, Y.G. Prasad, G.R. Chary, K.V. Rao, K.A. Gopinath, M. Osman, D.B.V. Ramana, M. Maheswari and V.U.M. Rao (2015). District Agriculture Contingency Plans to Address Weather Aberrations and for Sustainable Food Security in India. ICAR-Central Research Institute for Dryland Agriculture, Natural Resource Management Division, Hyderabad-500 059, India. 22p.

Prasad YG, Osman M, Singh SS, Manoranjan Kumar, Singh KM, Dixit S, Singh RD, Singh AK, Maheswari M, Bhatt BP, Venkateswarlu B, and Sikka AK. (2014). Contingency Measures for Deficit Rainfall Districts in South Bihar. CRIDA, Hyderabad. 19p.

Prasad YG et. al. (2015). Monsoon Action Plan - 2015: Village Level Contingency Plans for Climate Resilience in Agriculture. Technology Demonstration Component, National Innovations in Climate Resilient Agriculture (NICRA), ICAR-Central Research Institute for Dryland Agriculture, Hyderabad, 54p.

10

Soil Management for Climate-Resilient Agriculture

H. Pathak

Center for Environment Science and Climate Resilient Agriculture
Indian Agricultural Research Institute (IARI), New Delhi.

10.1 Introduction

Over the past few decades, the gaseous composition of the atmosphere has undergone a significant change mainly through increased industrial emissions, fossil fuel combustion, widespread deforestation and burning of biomass, as well as changes in land use and land management practices. These anthropogenic activities have resulted in an increased emission of naturally occurring radiatively active gases e.g., carbon dioxide (CO_2), methane (CH_4) and nitrous-oxide (N_2O), popularly known as the 'greenhouse gases'. These GHGs trap the outgoing infrared radiation from the earth's surface and thus raise the temperature. The global mean annual temperatures at the end of the 20th century, as a result of GHG accumulation in the atmosphere, has increased by 0.4–0.7 °C above that recorded at the end of the 19th century. The last 50 years show an increasing trend of 0.13 °C/decade while the trend of the last one and half decades has been much higher. The Inter-Governmental Panel on Climate Change (IPCC, 2014) projected a temperature increase between 3.7 and 4.8 °C by the end of the 21st Century. Global warming also leads to other regional and global changes in climate-related parameters such as rainfall, soil moisture, and sea level. The mean sea level has risen by 10–20 cm in last century. Similarly, snow cover is also believed to be gradually decreasing.

Soil contributes to the greenhouse effect and also acts as a sink of GHGs. It is intricately linked to the atmospheric–climate system through the carbon, nitrogen, and hydrologic cycles. This chapter assesses the contributions of soil to the emission of GHGs and its potential role in mitigation and adaptation to climate change. The article discusses the emission of greenhouse gases (GHGs) from agriculture and the strategies for mitigation and adaptation.

10.2 Soil as a Cause of Climate Change

Agriculture contributes to greenhouse effect primarily through the emission and consumption of GHGs such as CH_4, N_2O and CO_2. The CH_4 is produced in soil during microbial decomposition of organic matter under anaerobic conditions. Rice fields submerged under water are the potential source of CH_4 production. Continuous submergence, higher organic C content and use of organic manure in puddled soil enhance the CH_4 emission. Burning of crop residues also contributes to the global methane budget. The enteric fermentation in ruminants is another major source of CH_4 emission (IPCC, 2007).

Nitrous oxide is produced in soils through the processes of nitrification and denitrification. Nitrification is the aerobic microbial oxidation of ammonium to nitrate, and denitrification is the anaerobic microbial reduction of nitrate to nitrogen gas (N_2). Nitrous oxide is a gaseous intermediate in the reaction sequence of denitrification and a by-product of nitrification that leaks from microbial cells into the soil and ultimately into the atmosphere. One of the main controlling factors in this reaction is the availability of inorganic N in soil through additions of synthetic or organic fertilizers, manure, crop residues, sewage sludge or mineralization of N in soil organic matter following drainage/management of organic soils and cultivation/land-use change on mineral soils.

The main source of carbon dioxide in agriculture is the soil management practices such as tillage which triggers emission of this gas through biological decomposition of soil organic matter. Tillage breaks the soil aggregates, increases the oxygen supply and exposes the surface area of organic material promoting the decomposition of organic matter. Use of fuel for various agricultural operations and burning of crop residues are the other sources of carbon dioxide emission. An off-site source is the manufacturing of farm implements, fertilizers and pesticides.

India is a signatory to the United Nations Framework Convention on Climate Change (UNFCCC), which aims to stabilize GHG concentrations in the atmosphere at levels that would prevent dangerous anthropogenic interference with the climate system. The inventory of GHG emission for the year 2007 showed that Indian rice fields covering an area of 43.86 million ha emitted 3.37 million tons of CH_4 (Table 1). Total N_2O emission from agricultural soils of India was 0.22 million tons. Burning of crop residues in field emitted 0.25 million tons of CH_4 and 0.01 million tons of N_2O. Emission of methane from rice fields has remained constant over the years; however, emission of N_2O has increased, mainly because of increased N fertilizer use.

Table 10.1 Greenhouse gas emissions from Indian agriculture during 2007

Source	CH_4	N_2O	CO_2 eq.
		Million ton	
Enteric fermentation	10.10	-	212.09
Manure management	0.12	-	2.44
Rice cultivation	3.37	-	84.24
Agricultural soil	-	0.22	64.7
Crop residue burning	0.25	0.01	8.21
Total	**13.84**	**0.23**	**371.68**

Source: Pathak et al. (2010)

10.3 Impacts of Climate Change on Soil

Influence of global warming on soil processes is complex and affected by several direct as well as indirect factors. Increased precipitation will increase the surface runoff in hilly lands; increase infiltration and water storage within the soil in flat lands; enhance groundwater recharge in the highly permeable and well drained soils; increase the evaporation on soil having low infiltration and transpiration in the case of well-developed plant canopies. Rise in temperature will increase the potential evapo-transpiration and decrease the surface run off, infiltration, water-storage and groundwater recharge, especially if accompanied by low precipitation. The main impacts of climate change on soil are listed below.

- Reduced quantity and quality of organic matter content, which is already quite low in Indian soil.

- Under elevated CO_2 concentration, crop residues have higher C:N ratio, which may reduce their rate of decomposition and nutrient supply.

- Increase of soil temperature will increase N mineralization but its availability may decrease due to increased gaseous losses through processes such as volatilization and denitrification.

- Change in rainfall volume and frequency and wind intensity may alter the severity, frequency and extent of soil erosion.

- Rise in sea level may lead to salt-water ingression in the coastal lands turning them less suitable for conventional agriculture.

10.4 Greenhouse Gas Mitigation Strategies

The strategies for mitigating methane emission from rice cultivation could be altering water management, particularly promoting mid-season aeration by short-term drainage; improving organic matter management by promoting aerobic degradation through composting or incorporating it into soil during off-season drained period; use of rice cultivars with few unproductive tillers, high root oxidative activity and high harvest index; and application of fermented manure like biogas slurry in place of unfermented farmyard manure. Methane emission from ruminants can be reduced by altering the feed composition, either to reduce the percentage which is converted into methane or to improve the milk and meat yield.

The most efficient management practice to reduce nitrous oxide emission is site-specific nutrient management. The emission could also be reduced by nitrification inhibitors such as nitrapyrin and dicyandiamide (DCD). There are some plant-derived organics such as neem oil, neem cake and karanja seed extract which can also act as nitrification inhibitors.

Mitigation of CO_2 emission from agriculture can be achieved by increasing carbon sequestration in soil through manipulation of soil moisture and temperature, setting aside surplus agriculture land, and restoration of soil carbon on degraded land. Soil management practices such as reduced tillage, manuring, residue incorporation, improving soil biodiversity, micro aggregation, and mulching can play important roles in sequestering carbon in soil. The GHG mitigation potential of the most promising technologies and their constraints are summarized in Table 2. Some technologies such as intermittent drying and site-specific N management can be easily adopted by the farmers without extra investment whereas other technologies need economic incentives and policy support.

Table 10.2 Potential and constraints of greenhouse gas mitigation options

Option	Mitigation potential	Constraints
Methane from rice field		
Intermittent drying	25-30%	Assured irrigation
Direct-seeded rice	30-40%	Machine, herbicide
System of rice intensification	20-25%	Labour, assured irrigation
Methane from ruminants		
Balanced feeding	5-10%	Small holding
Feed additives	5-10%	Cost, biosafety
Nitrous oxide from soil		
Site-specific N management	15-20%	Awareness, fertilizer policy
Nitrification inhibitor	10-40%	Cost, incentive

Source: Pathak et al. (2010)

10.5 Low Carbon Technologies in Agriculture

Scientific agriculture can help in mitigating GHGs emission. The following strategies have been recommended for mitigating methane emission from rice cultivation.

- altering water management, particularly promoting intermittent irrigation and mid-season drainage;

- improving organic matter management by promoting aerobic degradation through composting or incorporating it into soil during off-season drained period;

- use of rice cultivars with few unproductive tillers, high root oxidative activity and high harvest index; and

- application of fermented manure like biogas slurry in place of unfermented farmyard manure (Pathak et al., 2010). A single mid-season drainage may reduce seasonal methane emission. This emission could be reduced further by intermittent irrigation, yielding a 30% reduction as compared to mid-season drainage.

Emission of N_2O can be reduced by following management practices that improve N-use efficiency including using slow or controlled release of fertilizer or nitrification inhibitors which retard the microbial processes leading to N_2O formation. The most efficient management practices to reduce nitrous oxide emission are:

- site-specific nutrient management, and

- use of nitrification inhibitors such as nitrapyrin and dicyandiamide.

There are some plant-derived organics such as neem oil, neem cake and karanja seed extract which can also act as nitrification inhibitors. Nitrification inhibitors reduce N_2O emission directly by reducing nitrification, and indirectly by reducing the availability of NO_3 for denitrification.

Twenty technologies have been analysed for their potential to mitigate GHGs emission in rice in the upper and lower Indo-Gangetic Plains (IGP) and their economic feasibilities have been assessed (Pathak et al., 2012). During crop production under conventional management practices, global warming potential (GWP) of rice cultivation was 3957 kg CO_2 ha^{-1} in the upper-IGP and 2934 kg CO_2 ha^{-1} in the lower-IGP. Compared to the current practices of farmers, 15 technologies in the upper-IGP and 14 technologies in the lower-IGP have the potential to reduce the GWP. In the upper-IGP, only seven technologies, viz. sprinkler irrigation, direct seeded rice, use of nitrification inhibitor, use of urea super granules, use of leaf colour chart, site-specific nutrient management and crop diversification have depicted ability to reduce GWP without any additional cost. In the lower-IGP, use of nitrification inhibitor, use of leaf colour chart, site-specific nutrient management and crop diversification have shown reduction in GWP with no additional cost.

The major benefits of low carbon agricultural technologies are savings in irrigation water, labour and energy; reduction in GHGs emission; better water- and nutrient-use efficiency; provision of tolerance to moisture and heat stresses; improvement in soil health; and increase in income. There are some constraints also; these include high initial cost, infrastructure for installation and maintenance, knowledge-intensiveness and technical capability, high production cost, risk in rainfed areas, weed problem, yield loss, inadequate market facility, lack of awareness and limited post-harvest facilities.

The policy interventions required to overcome the constraints are development of irrigation facilities, incentives for saving of water, carbon credits for mitigation, subsidy and incentive for installation of resource conserving infrastructure, trainings to farmers for skill development, public awareness generation, development of effective, low-cost, environment-friendly herbicides, accurate weather forecasting, development of post-harvest facilities and refining of technologies to make them simple, cheap and effective.

Several options are available for mitigating GHGs emission in agriculture. Policies and incentives would have to be developed to encourage farmers for adopting these mitigation options to harness benefits of improved soil health and better water- and energy-use efficiencies. Benefits, challenges and required interventions for promoting low-carbon and cost-effective mitigation technologies in rice and wheat in the Indo-Gangetic Plains are given in Table 10.3.

Table 10.3 Interventions for promoting mitigation technologies in rice and wheat in the Indo-Gangetic Plains.

Technology	Major benefits[a]	Key challenges	Required interventions
Rice			
Alternate wetting and drying/ Mid-season drainage	• Saving in irrigation water (25-30%) • Saving in labor (10-15%) and energy • Reduction in methane emission (70-75%)	• Weed problem • Yield loss (5-10%) • Difficulties and risk in rainfed areas • Increase in N_2O emission (5-10%)	• Development of effective, low cost, environment friendly herbicides • Accurate weather forecasting • Development of Irrigation facilities • Provision of incentives for saving water • Introduction of carbon credit for mitigation
Drip and sprinkler irrigation	• Saving in irrigation water (25-30%) • Saving in energy • Higher water- and nutrient-use efficiencies	• High initial cost • Need of infrastructure for installation and maintenance • Highly technical	• Introduction of subsidy for installation • Organization of trainings for skill development • Generation of public awareness • Introduction of carbon credit for mitigation
Nitrification inhibitor/ urea super granules	• Reduction in losses and increase in N-use efficiency	• High production cost • Technical application process • Non-availability in the market	• Motivation of industries to manufacture these products • Provision of incentives to farmers • Generation of public awareness • Introduction of carbon credit for mitigation

Table 10.3 *Contd...*

Technology	Major benefits[a]	Key challenges	Required interventions
Site-specific N management/ use of LCC	• Reduction in losses and increase in N-use efficiency	• Lack of awareness • Soil testing facility inadequate	• Establishment of more soil testing laboratories • Provision of incentives to farmers • Generation of public awareness • Introduction of carbon credit for mitigation
Direct seeding of rice	• Saving in irrigation water (25-30%) • Saving in labor (10-15%) and energy • Reduction in methane emission (70-75%) • Increase in stress tolerance	• Knowledge intensive • Weed problem • Yield loss (5-10%) • Increase in N_2O emission (5-10%)	• Development of effective, low cost, environment-friendly herbicides • Training to farmers • Provision of incentive for saving of water • Introduction of carbon credit for mitigation
Crop diversification (rice to maize)	• Saving in irrigation water (50-60%) • Saving in labor (20-25%) and energy • Reduction in methane emission (90-100%)	• Minimum support price not available • Inadequate market • Lack of awareness • Limited post-harvest facilities	• Provision of minimum support price • Development of market infra-structure • Awareness generation • Development of post-harvest facilities
Wheat			
Zero tillage	• Saving in fuel (70-90 L diesel ha^{-1}) • Saving in water	• Knowledge intensive • Non-availability and maintenance of quality zero-till drill • Lack of awareness	• Refinement of technology • Skill development for operation and maintenance of drill • Awareness generation

Table 10.3 *Contd...*

Technology	Major benefits[a]	Key challenges	Required interventions
Integrated nutrient managemen t/organic farming	• Improvement in soil health • Increase in N-use efficiency • Reduction in GWP • Increase in income	• Non-availability of manure • Poor quality of manure • High transportation cost	• Promotion of agri-wastes for compost • Promotion of biogas for energy and manure • Introduction of carbon credit for biogas
Nitrification inhibitor/Sit e-specific N managemen t	• Reduction in losses and increase in N-use efficiency	• High production cost • Technical application process • Non-availability in the market	• Motivation of industries to manufacture these products • Provision of incentives to farmers • Generation of public awareness • Introduction of carbon credit for mitigation

Note: Figures within the parenthesis are compared to conventional practices

Source: Pathak et al. (2012)

10.6 Climate-smart Soil Management Practices

Climate-smart soil management practices includes, besides others, precision land leveling, no-till systems, direct seeded rice, bed planting, crop residue management, crop diversification, and integrated nutrient management (INM)/site specific nutrient management.

10.6.1 Laser-aided Land Leveling

The laser-aided precision land leveling, so-called laser land leveling (LLL) is a precursor technology for adopting conservation agriculture (CA) practices like zero tillage, bed planting. Traditionally leveled farmers' field have 5-15 cm undulation in general within the field, which often results in poor seedling emergence, seedling mortality due to water logging and uneven initial crop growth. Laser land leveling enables the farmers to apply water and nutrient uniformly facilitating a uniform crop stand and maturity through improved nutrient-water interactions. Results revealed that the adoption of laser leveling technology is accelerating at multiple rates and currently nearly 0.2 M ha area has been brought under this technology in the IGP and saved significant energy, irrigation water, fuel, and electricity in addition to the yield advantages in several crops and cropping system. The mean water productivity at Ludhiana, Punjab in wheat under LLL was 0.52 kg grain/m^3 compared with 0.33 kg/m^3 under unlevelled condition.

10.6.2 No-till/Zero-tillage System

In no-till/zero till system, the seed is placed into soil by a seed drill without prior land preparation. In India, the burning of non-conventional fuel and resultant emission of GHGs is severe in the Indo-Gangetic basin. Rice-wheat is the dominant system in this region, where conventional method of land preparation/sowing not only disturbs the soil environment but also leads to atmospheric pollution. No-tillage systems eliminate all pre-planting mechanical seedbed preparation thereby reducing the fossils fuel burning and reduces sequester the carbon in the soil that ultimately reduces GHGs emissions. For controlling weeds, non-selective herbicides like paraquat, glyphosate may be applied at 1.0 kg/ha at 7-10 days before sowing of seed, followed by crop-selective herbicides in standing crops. Crop/plant residues/ stubbles or other vegetative substances should be retained on soil surface under zero-till conditions to reduce compaction of soil. This will provide favorable effect on crop root development and subsequent growth. Zero tillage leads to 5-10% yield advantage and reduction in cost of cultivation. Farmers cultivating fields 3 to 4 times before sowing of crops, resulting in extra-cost spent on fossil fuel can be avoided by adopting zero tillage. Nearly Rs. 2500-3000/ha can be saved. Besides, wear and tear losses of tractor and machinery are minimized. Water requirement is reduced. Weed infestation is also reduced over time. The loss of organic carbon by oxidation due to conventional tillage is reduced, the activity of micro-flora and fauna is increased with residues, and the production potential of soil is enhanced through zero tillage. In zero-till cultivation, proper soil moisture should be present in soil at the time of sowing.

10.6.3 Direct-seeded Rice

Aerobic rice culture/direct-seeded rice has great potential for minimizing the cost of production, soil health hazards and the negative impacts on the succeeding crops. Further, reduction in CH_4 emission from rice-paddies is observed due to adoption of resource conservation technologies (aerobic rice), particularly in rice-wheat production system. Direct seeding is another complement to conservation agriculture. Although transplanting of crops, including rice, is possible under zero tillage, direct seeding is preferable for the reasons mentioned above. In addition direct seeding results in less soil movement than transplanting, which often involves some sort of strip tillage. Under the changing agricultural scenario, the agricultural technologies needs a shift from production oriented to profit oriented sustainable farming. In this direction, the pace of adoption of resource conserving technologies (RCTs) by the Indian farmers is satisfactory to a larger extent, but under the present scenario, we are in the half way of conservation agriculture. The CA systems will lead to sustainable farming and will be the most thrust of the future farming.

Direct-seeded rice (DSR) avoids water required for land preparation/puddling and reduces overall water demand compared to puddled transplanted rice (TPR). It is a labour, fuel, time, and water-saving technology and gives similar yield to TPR, if weeds were controlled with judicious use of herbicides. DSR does not affect rice quality and can be practised in different ecologies, including upland, medium and lowland, deep water and irrigated areas by large as well as small farmers. Soil health is maintained or improved, and fertilizer and water use efficiencies are higher in DSR (saving of 30-40% irrigation water). Therefore, DSR is technically and economically a feasible alternative to TPR. The wet season DSR should be planted 10-12 days before the historical trends for the onset of monsoon. In North Indian conditions, summer mungbean can be adopted without delay in sowing of rice crop. It gives grain yield of 8-10 q ha^{-1} and usually adds 40-60 kg N ha^{-1} in soil, reducing requirement of N of subsequent crop.

10.6.4 System of Rice Intensification (SRI)

It is a cultivation practice for rice that is taken up in a different and more biologically enriched environment for growth. Yields are increased by 50-100% or more, with a reduction in plant populations (by 80–90%), less water (by 25-50%), without using new 'improved' varieties (all varieties respond to the methods) or using chemical fertilizers (just adding compost to the soil), with usually lowered costs of production, and thus considerably increased net economic returns per hectare.

10.6.5 Bed Planting

It is a system in which crops are sown on ridges or beds. The height of the beds is maintained at about 15 to 20 cm and having a width of about 40 to 70 cm depending on the crops. In case of wheat around 45 cm bed width is maintained and generally three rows having a distance of 15 cm are sown. The furrow width is generally 25 cm. During the last decade practice of raised bed planting has emerged with a greater pace in the IGP. The major concern of this system is to enhance the productivity and save the irrigation water.

Potential agronomic advantages of beds include improved soil structure due to reduced compaction through controlled trafficking, and reduced water logging and timely machinery operations due to better surface drainage. Permanent beds conserve moisture during prolong dry spells in water scarcity areas for longer availability of moisture to sustain plant life and acts as a drainage channel in high rainfall areas. The beds also create the opportunity for mechanical weed control and improved fertilizer placement. In rice-wheat systems in Asia and Australia permanent beds also provide the opportunity for diversification to water logging sensitive crops not suited to conventional flat sowing. Typical irrigation savings range from 18% to 30-50% in bed planted crops (Jat et al., 2005).

10.6.6 Crop Residue Management

Permanent or semi-permanent crop residue cover on soil, which can be live or dead mulch, has a role to protect soil physically from sun, rain and wind and to feed soil biota that take over the tillage function and nutrient balancing. Crop residue retention under zero-tillage is more beneficial than residue incorporation in conventional tillage. It reduces weeds, conserves soil moisture and organic carbon, regulates soil temperature and supplies nutrients, which ultimately reduces irrigation need and increases crop yield and net returns. Crop residues significantly influence physical, chemical and biological properties of soil and help in water conservation through enhanced water infiltration, and reducing evaporation, and wind and water erosion. Proper residue management will avoid straw burning, improve soil organic C, enhance input-use efficiency and have the potential to reduce GHGs emissions. New variants of zero-till seed-cum-fertilizer drill such as happy seeder, turbo seeder and rotary-disc drill can do direct drilling of seeds in the presence of surface residues (loose and anchored up to 10 t/ha). Retention of crop residues as mulch raises the minimum soil temperature in winter due to reduction in upward heat flux from soil and decreases soil temperature during summer due to shading effect. The residue retention on the soil surface slows the runoff by acting as tiny dams, reduces surface crust formation and enhances infiltration. The channels (macropores) created by earthworms and old plant roots, when left intact with no-till, improve infiltration to help reduce or eliminate runoff. Reduced evaporation from the upper strata of soil coupled with improved soil characteristics essentially leads to higher crop yield in many cropping and climatic situations.

10.6.7 Crop Diversification and Rotation

Crop diversification proved to be of paramount importance in mitigating the environmental problems arising on account of monoculture. Inclusion of certain crops in sequential and intercropping systems has been found to reduce some obnoxious weeds to a considerable extent, thereby reducing herbicides needs. Nitrate leaching is inevitable under most agricultural production systems. Choice of appropriate cropping systems and management practices helped minimizing nitrate leaching besides improving N-use efficiency. Inclusion of legumes in cropping systems has been found to be effective in reducing the nitrate leaching in lower profiles. Crop rotations also serve different purposes in the system and are linked to the other two principles. Besides, the phyto-sanitary and weed management objectives, crop rotations serve to open different soil horizons with different rooting types. The crop rotation becomes the part of the soil cover and residue management strategy with the objective to keep the soil constantly covered either under a live crop or dead residue mulch.

10.6.8 Integrated Nutrient Management (INM)

Since CH_4 and N_2O have GWPs so much higher than CO_2 (21 and 310 times more, respectively), agricultural management that reduces the emissions of these two GHGs is likely to have a large mitigation potential. Improved cropping system and nutrient management practices are needed to increase crop yield and quality to meet growing global food, fiber, and biofuel. Nutrient and cropping system management should also consider opportunities for reduced CH_4 and N_2O emissions especially per unit of food, fiber, or biofuel produced. Integrated nutrient management (INM) or integrated nutrient supply (INS) system aims at achieving efficient use of chemical fertilizers in conjunction with organic manures for reducing the green house gas emission. Long term fertilizer experiments involving intensive cereal based cropping systems reveal a declining trend in productivity even with the application of recommended levels of N, P and K fertilizers (Mahajan and Sharma, 2005). Conservation agriculture improves the soil aggregate and aggregate-associated N in the Indo-Gangetic plains under irrigated condition (Bhattacharyya et al., 2013). The crop productivity increases from the combined application of chemical fertilizers and organic manures. Such combination contributed to the improvement of physical, chemical and biological properties and soil organic matter and nutrient status. Integrated nutrient management is based on 4R Nutrient Stewardship Promotion by IPNI like i) right source of nutrient, ii) right fertilizer rate, iii) right timing of application and iv) right method of placement. In the past research mineral fertilizers boosted the rice-wheat yield in the IGP, but their rising cost and diminishing resources led to search for alternative sources of plant nutrients. Complementary use of the available renewable sources of plant nutrients (Organic/biological) along with mineral fertilizers led to the development of INM system.

10.6.9 Site Specific Nutrient Management (SSNM)

Fertilizer nitrogen is one of the major inputs in crop production. As fertilizer N has generally been managed following blanket recommendations consisting of two or three split applications of preset rates of the total amount of N, improvement in N use efficiency could not be achieved beyond a limit. Feeding crop N needs is the most appropriate fertilizer N management strategy to further improve N use efficiency. Since plant growth reflects the total N supply from all sources, plant N status at any given time should be a better indicator of the N availability. The chlorophyll meter (SPAD meter), Green seeker and leaf colour chart have emerged as diagnostic tools, which can indirectly estimate crop N status of the growing crops. LCC is an ideal tool to optimize N use in Rice/Maize at high yield levels, irrespective of the source of N applied, *viz.*, organic manure, biologically fixed N, or chemical fertilizers. Thus, it is an eco-friendly tool in the hands of farmers. Now, it is manufactured with 4 colors called Four Panel LCC & 6 colors called Six Panel LCC. Moreover, LCC is provided with water-proof laminated instruction sticker in the required regional language. Nitrogen management through LCC in hybrid as well as inbred rice has been found superior to local recommendation of N application in three splits, and it curtailed 20-30 kg of fertilizer N/ha without sacrificing rice yield. Nitrogen application at LCC<3 in Basmati 370 and at LCC<4 in coarse and hybrid rice was found optimum. Moreover, in LCC-based N management, basal application of N can be skipped without any disadvantage in terms of grain yield, and agronomic, physiological or recovery efficiency of fertilizer N.

10.7 Conclusion

The climate change may considerably affect the food supply and access through direct and indirect effects on crops, soils, livestock, fisheries and pests. There are, however, ways by which the adverse impacts may be mitigated and agriculture can be adapted to the changed climate. Soil management offers promises for climate change adaptation through modifying crop management practices, improving water management, adopting new farm techniques such as resource conserving technologies (RCTs), crop diversification and harnessing the indigenous technical knowledge of farmers. A win-win solution is to start with such mitigation strategies that are needed for sustainable development such as increasing soil organic C content. There is a need to develop policy framework for implementing the mitigation options so that the farmers are saved from the adverse impacts of climate change. Development of technologies for mitigation and their uptake at speedy rate by the farmers are essential for climate change management. Development and operationalization of adaptation strategy necessitate socio-psychological empowerment of farmers besides developing competencies in acquiring knowledge and skills related to adaptation practices.

References

Bhattacharyya, Ranjan, Das, T. K., Pramanik, P., Ganeshan, V., Saad, A. A. and Sharma, A. R. (2013). Impacts of Conservation Agriculture on Soil Aggregation and Aggregate-associated N under an Irrigated Agro-ecosystem of the Indo-Gangetic Plains. *Nutrient Cycling in Agroecosystems* 96: 185–202p.

INCCA (Indian Network for Climate Change Assessment). (2010). Assessment of the Greenhouse Gas Emission: 2007. The Ministry of Environment & Forests, Govt. of India, New Delhi.

IPCC (Inter-governmental Panel on Climate Change). (2007). The Physical Science Basis. In: Solomon, S, Qin D, Manning M, Chen Z, Marquis M, Averyt KB, Tignor M, Miller HL (eds.) Climate Change 2007: Contribution of Working Group I to the Fourth Assessment Report of the IPCC. Cambridge University Press, Cambridge, UK.

Jat, M.L., Gathala, M.K., Ladha, J.K., Saharawat, Y.S., Jat, A.S., Kumar, V., Sharma, S.K., Kumar, V. and Gupta, R.K. (2009). Evaluation of

Precision Land Leveling and Double Zero till Systems in Rice-Wheat Rotation: Water use, Productivity, Profitability and Soil Physical Properties. *Soil and Tillage Research* 105: 112-121p.

Pathak H and Aggarwal PK (Eds.) (2012). Low Carbon Technologies for Agriculture: A Study on Rice and Wheat Systems in the Indo-Gangetic Plains. Indian Agricultural Research Institute, 76p.

Pathak H, Bhatia A, Jain N and Aggarwal PK (2010). Greenhouse Gas Emission and Mitigation in Indian Agriculture – A review. In: ING Bulletins on Regional Assessment of Reactive Nitrogen, Bulletin No. 19, Ed: Bijay Singh. SCON-ING, New Delhi.

11

Crop Specific Technologies for Climate Resilient Agriculture

J. V. N. S. Prasad

*ICAR- Central Research Institute for
Dryland Agriculture (CRIDA),Hyderabad.*

11.1 Introduction

The Intergovernmental Panel on Climate Change in its Fifth Assessment Report has clearly brought out that the warming of the climate system is unequivocal and climate change is happening. The atmosphere and oceans have warmed, the amount of snow and ice have diminished, sea level has risen, and the concentrations of greenhouse gases have increased. The globally averaged combined land and ocean surface temperature data show a warming of 0.85 °C, over the period 1880 to 2012 (IPCC, 2013). The warming will have widespread impacts on human health, settlements and natural resources, if no measures are taken to curb the ill-effects of global warming. It observed that the coming years will see more extreme weather events (floods, cyclones, cloud bursts, unseasonal excessive rains and drought, etc.) in most parts of the globe and India will be among the most affected countries.

11.2 Climate Change and Indian Agriculture

Climate change impacts on agriculture are being witnessed all over the world, but countries like India are more vulnerable in view of the high population depending on agriculture and excessive pressure on natural resources. The warming trend in India over the past 100 years (1901 to 2007) was observed to be 0.51 °C with accelerated warming of 0.21 °C per every10 years since 1970 (Krishna Kumar 2009). The projected impacts are likely to further aggravate yield fluctuations of many crops with impact on food security and prices. Cereal productivity is projected to decrease by 10-40% by 2100 and greater loss is expected in *rabi*. There are already evidences of negative impacts on yield of wheat and paddy in parts of India due to increased temperature, increasing water stress and reduction in number of rainy days. Modeling studies project a significant decrease in cereal production by the end of this century. Climate change impacts are likely to vary in different parts of the country. Parts of western Rajasthan, Southern Gujarat, Madhya Pradesh, Maharashtra, Northern Karnataka, Northern Andhra Pradesh, and Southern Bihar are likely to be more vulnerable in terms of extreme events (Mall *et al.* 2006). For every one degree increase in temperature, yields of wheat, soybean, mustard, groundnut and potato are expected to decline by 3-7% (Agarwal 2009). Similarly, rice yields may decline by 6% for every one degree increase in temperature (Saseendran *et al.* 2000). Water requirement of crops is also likely to go up with projected warming and extreme events are likely to increase.

Rainfed agriculture is likely to be more vulnerable in view of its high dependency on monsoon, the likelihood of increased extreme weather events due to aberrant behavior of south west monsoon. Nearly 85 m ha of India's 141 m ha net sown area is rainfed. Rainfed farming area falls mainly in arid, semi-arid and dry sub-humid zones. About 74% of annual rainfall occurs during southwest monsoon (June to September). This rainfall exhibits high coefficient of variation particularly in arid and dry semi-arid regions. Skewed distribution has now become more common with reduction in number of rainy days. Aberrations in South-West monsoon which include delay in onset, long dry spells and early withdrawal, all of which affect the crops, strongly influence the productivity levels (Lal 2001).

11.3 Observed Trends in
Key Weather Parameters and Crop Impacts

Rainfall is the key variable influencing crop productivity in rainfed farming. Intermittent and prolonged droughts are a major cause of yield reduction in most crops. Long term data for India indicates that rainfed areas witness 3-4 drought years in every 10-year period. Of these, 2-3 are in moderate and one may be of severe intensity. A long term analysis of rainfall trends in India (1901 to 2004) using Mann Kendall test of significance by AICRPAM-CRIDA indicate significant increase in rainfall trends in West Bengal, Central India, coastal regions, south western Andhra Pradesh and central Tamil Nadu. Significant decreasing trend was observed in central part of Jammu Kashmir, Northern MP, Central and western part of UP, northern and central part of Chattisgarh (Figure 11.1). Analysis of number of rainy days based on the IMD grid data from 1957 to 2007 showed declining trends in Chattisgarh, Madhya Pradesh, and Jammu Kashmir. In Chattisgarh and eastern Madhya Pradesh, both rainfall and number of rainy days are declining which is a cause of concern as this is a rainfed rice production system supporting large tribal population who have poor adaptive capabilities.

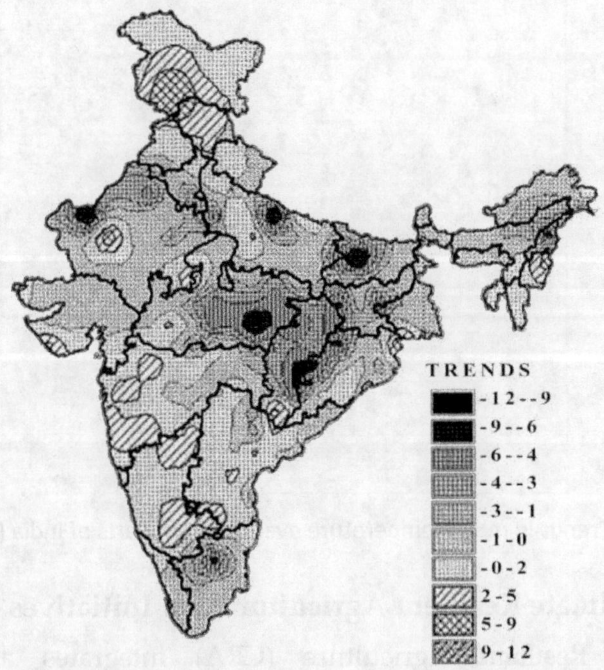

Fig. 11.1 Rainfall trends over India from 1901 to 2004 (NPCC,2007)

Temperature is another important variable influencing crop production particularly during *rabi* season. An analysis carried out by AICPRAM-CRIDA using maximum and minimum temperature data for 47 stations across India showed that 9 out of 12 locations in south zone showed an increasing trend for maximum temperature, where as the north, only 20% locations showed increasing trend (Figure 11.2). For minimum temperature most of the stations in India are showing increasing trend. This is a cause of concern for agriculture as increased night temperatures accelerate respiration, hasten crop maturity and reduce yields. The increasing trend is more evident in central and eastern zones where rainfall is also showing a declining trend which is an area of concern and requires high attention for adaptation research.

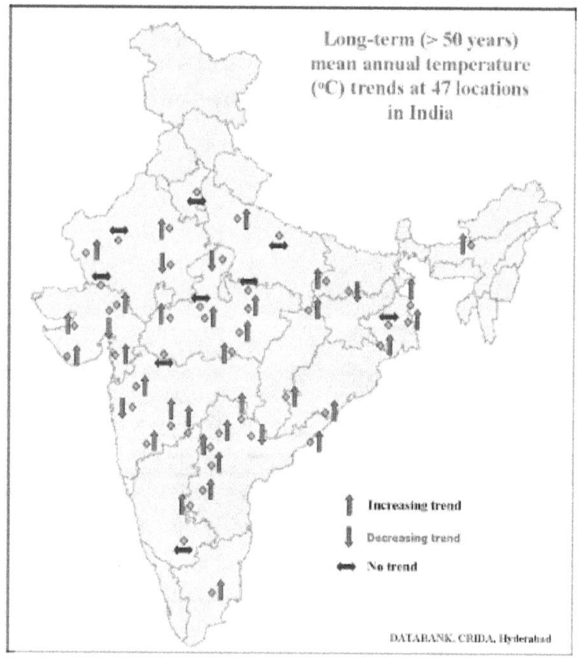

Fig. 11.2 Trends in mean temperature over different parts of India (NPCC-2009)

11.4 Climate Resilient Agriculture and Initiatives by ICAR

Climate Resilient Agriculture (CRA) integrates adaptation, mitigation and other practices in agriculture which increases the capacity of the system to respond to various climate related disturbances by resisting damage and recovering quickly. Such perturbations and disturbances can include events such as drought, flooding, heat/cold wave, erratic rainfall pattern, long dry spells, insect or pest population explosions and other perceived threats caused by changing climate. In short it is the ability of the system to bounce back. Climate resilient agriculture includes an in-built property in the system for the recognition of a threat that needs to be responded to, and also the degree of effectiveness of the response. CRA will essentially involve judicious and improved management of natural resources viz., land, water, soil and genetic resources through adoption of best bet practices.

The Indian Council of Agricultural Research (ICAR) has launched the National Initiative on Climate Resilient Agriculture (NICRA), a comprehensive project covering strategic research, technology demonstration and capacity building. The objectives of NICRA are to enhance the resilience of Indian agriculture covering crops, livestock and fisheries to climatic variability and climate change through development and application of improved production and risk management technologies; to demonstrate site specific technology packages on farmers' fields for adapting to current climate risks; to enhance the capacity building of scientists and other stakeholders in climate resilient agricultural research and its application (Prasad et al., 2015).

The specific objectives of technology demonstration component are to demonstrate site specific technology interventions on farmers fields for coping with climate variability in vulnerable districts; to generate awareness and build capacity among farmers and other stakeholders on climate resilient agriculture; to evolve innovative institutional mechanisms at village level that enable the communities to respond to climatic stresses.

As part of the technology demonstration component, 100 climatically vulnerable districts were identified based on a scientific analysis of climate related problems, farmers' experiences and perceptions following a bottom-up approach. One village or a cluster of villages from each of the 100 selected districts was selected for this purpose by the respective Krishi Vigyan Kendra (KVK), in the district. Planning, coordination and monitoring of the program at national level is the responsibility of CRIDA. Eight Zonal Project Directorates (ZPDs) are involved in coordinating the project in their respective zones. At districts level, the selected KVK is responsible for implementing the project at village level through farmer's participatory approach. About 15 KVKs are involved in the states of Andhra Pradesh, Telangana and Karnataka which are vulnerable to the climate variability such as drought, flood and cyclone. Various practices are being demonstrated for reducing the vulnerability and to enhance the adaptive capacity and some of the proven crop specific technologies for these regions are given below:

11.5 Improved Varieties which can Escape Drought and Flood

Some of the varieties with significantly higher production potential can mature relatively early in comparison to the commonly grown varieties and can escape drought at maturity thus avoiding the yield loss due to early withdrawal of monsoon. In recent years, a number of varieties were released by the National Agricultural Research System which can mature relatively early and can escape drought. A comprehensive list of drought tolerant varieties suitable for these regions in various crops is given by Maheswari et al., (2015). These varieties can escape drought like situations in the event of early withdrawal of monsoon. Many flood tolerant varieties were released and some of these varieties can tolerate prolonged flooding. Such varieties can minimize the loss due to the flooding during the cropping season.

Flood tolerant varieties: In Sirusuwada village of Srikakulam district which is frequently affected by floods, flood tolerant paddy varieties were compared with that of the farmers' practice of Swarna and Pooja varieties. Of all the varieties tested, RGL-2537 recorded higher grain yield followed by MTU-1061 in low inundation area where as in high and medium inundation areas all the flood tolerant varieties were affected by flood (Table 11.1 and Figure 11.3& 11.4).

Table 11.1 Performance of flood tolerant paddy varieties in
Srikakulam district of Andhra Pradesh

Particulars	MTU-1061	MTU-1064	RGL-2537	PLA-1100	SWARNA	POOJA
Average yield (q/ha)	54	52	58	51.4	47.5	48.4
Cost of cultivation (Rs/ha)	38750	38750	38750	38750	38750	38750
Gross returns (Rs/ha) @ Rs. 1280/q	69120	66560	74240	65792	60800	61952
Net returns (Rs/ha)	30370	27810	35490	27042	22050	23202
B: C ratio	1.78:1	1.71:1	1.91:1	1.69:1	1.56:1	1.59:1

Fig. 11.3 Flood tolerant paddy varieties (PLA-1100, RGL-2537) at Srikakulam, AP

In Undi village of west Godavari, paddy varieties were evaluated for their flood tolerance with MTU-1061, MTU 1064 paddy varieties which are submergence and lodging tolerant along with the farmers' practice of swarna (MTU-7029). During October 2014 there was hudhud cyclone in Andhra Pradesh and it was observed that the MTU 7029 (Swarna) variety was completely affected with BPH after hudhdud cyclone. Whereas the MTU-1061 & MTU-1064 were less affected with BPH.

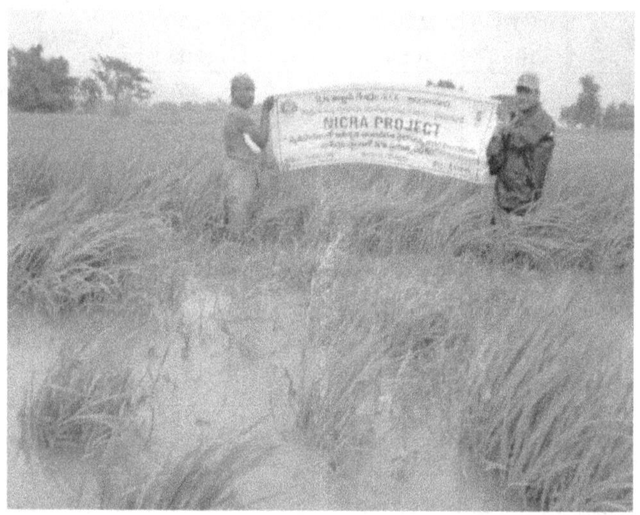

Fig. 11.4 Flood affected paddy (Pooja) at Srikakulam, AP

Table 11.2 Performance of flood tolerant paddy varieties in West Godavari district of Andhra Pradesh

Treatments	Seed yield (kg/ha)	Percentage of BPH incidence	Cost of cultivation (Rs/ha)	Gross income (Rs/ha)	Net income (Rs/ha)	B:C ratio
MTU-7029 (Common variety)	5247	75%	45357	66857	21500	1.47
MTU-1061 (Submergence & lodging tolerant variety)	5975	50%	40892	76867	35975	1.87
MTU1064 (Submergence & lodging tolerant variety)	5905	45%	39462	81112	41650	2.41

11.6 Improved Methods of Paddy Cultivation

In recent years various new methods of rice cultivation are being developed. Some of them are SRI (system of rice intensification), aerobic rice, direct seeded rice, etc., which are being practiced in various parts of the country. The water management practices in these systems is variable. The traditional method of paddy cultivation causes continuous ponding of water leading to prolonged anaerobic condition resulting in the emissions of methane. Any method of rice cultivation which results in reduction in the duration of submergence reduces the methane emissions. New methods of rice cultivation such as aerobic rice saves water to the extent of 55-60%, SRI method of cultivation results in saving of water to the extent of 25-30% where as alternate wetting and drying saves about 20%. Some of these methods aim at saturation and not submergence and experimental results show that continuous maintenance of saturation reduces methane emissions in comparison to the continuous ponding of water.

In Khammam district of Telangana, significant area was affected by salinity and KVK has introduced a salt tolerant paddy cultivar (WGL-44), which has yielded higher over the farmer's practice.

Table 11.3 Performance of Sidhi (WGL-44) paddy variety in Khammam district of Andhra Pradesh

Intervention	Technology demonstrated	No. of farmers	Area (ha)	yield[*] (kg/ha)	B:C Ratio
Salinity tolerant paddy variety (Siddhi-WGL-44)	Siddhi-WGL-44	6	0.2	6120	1:1.4

11.7 Residue Recycling and Zero Till Sowing of Maize in Paddy Fallows

In paddy-fallow systems in Andhra Pradesh, Telangana and Karnataka where paddy straw is commonly burnt after the harvest of the paddy, zero till planting is one of the important practice. Zero-tillage refers to direct drilling of maize in unploughed paddy fields immediately after rice harvest using zero till drill or happy seeder. It has been proven that zero till maize can result in higher yields, result in savings in irrigation water and advancing the sowing of maize by a fortnight. Recycling of paddy residues and zero till maize has multiple advantages; viz. reduces the CO_2 emissions into the environment by avoiding of burning of paddy residues, reduces the soil disturbance and savings in water and energy for pumping.

In Sirsuwada village of Srikakulam, rice-rice and rice-pulse cropping systems are practiced. Blackgram taken up in paddy fallows suffers from yellow mosaic virus disease and exposure to low temperature and fog at the time of flowering resulting in low yields. Maize is a possible alternative after rice. Zero till maize was adopted by 15 farmers in 5 ha area saved 2 irrigations and also reduced cost of land preparation (Figure 11.5).

Fig. 11.5 Zero till maize in rice fallows at Srikakulam, AP

11.8 Site Specific Nutrient Management

Integrated Nutrient Management and Site-Specific Nutrient Management (SSNM) is another approach with potential to climate change mitigation. Demonstrated benefits of these technologies are; increased rice yields and thereby increased CO_2 net assimilation and 30-40% increase in nitrogen use efficiency and decreasing GHG emissions linked with N fertilizer use in rice systems. Application of soil test based fertilizer helps to minimize the excess use of a nutrient and helps to identify deficient nutrients and optimize the nutrient application. This practice helps in enhancing the productivity as it ensures the availability of limiting nutrients and also minimizes the excess application of nutrients. Phosphorus (P) deficiency, for example, not only decreases yields, but also triggers high root exudation and increases CH_4 emissions. Judicious fertilizer application, a principal component of SSNM approach, thus has twofold benefit, i.e., reducing greenhouse gas emissions; at the same time improving yields under high CO_2 levels. The application of a urease inhibitor, hydroquinone (HQ), and a nitrification inhibitor, dicyandiamide (DCD) together with urea also is an effective technology for reducing N_2O and CH_4 from paddy fields.

In Nalgonda, soil test based nutrient management was taken up in cotton. About 30 soil samples were collected from Nandhyala gudem, Atmakur mandal and all the soil samples are low in organic content. The available nitrogen is low in all the samples i.e., less than 280 kg/ha and the available P_2O_5 is low to medium and available potash is low to medium. Recommended fertilizer dosage was 80:24:24 NPK kg/ha as against farmers' practice of 120:60:50 NPK kg/ha. Farmers could save Rs.1200 - 1500/ha towards the cost of fertilizers due to soil test based nutrient application.

11.9 Introducing Varieties which are Resistant to Prevalent Diseases of the Region

Some of the regions are endemic for certain diseases and pests and spread of these pests is aggravated by favourable weather resulting in significant crop losses. In those regions, a high yielding variety which is tolerant/ resistant to a particular pest needs to be introduced. For example: in some of the chickpea grown regions wilt is of common occurrence and Yellow vein mosaic virus is a serious problem in some of the blackgram/ green gram/ soybean growing regions. White fly in cotton is of regular occurrence in some of the regions. The problems in the target district/ region with reference to the predominant crops are to be ascertained and accordingly suitable tolerant varieties needs to be introduced.

11.10 Improved Intercropping Systems

Intercropping systems with a combination of short duration and a long duration crop, a shallow rooted with a deep rooted crop, a legume and a non legume crop is a resilient practice as the system can perform better in the event of variable rainfall. In the event of the early season drought, the short duration crop may get affected where as the long duration crop can perform better. In the event of the terminal drought, the short duration crop will perform better and provide farmer the normal yields and thus imparts resilience to the system. Similarly by combining the legume and non legume crops, the synergies and the complimentarity between the crops can be better harnessed. The introduced intercropping systems should have components that meet the requirements of the household.

11.11 Crop Diversification

Diversifying with crops which gives assured income with assured market and provide high returns to the farming community are to be explored. The criterion of selection of new crop should be such that the alternate crops is more tolerant to the moisture stress or should withstand the rainfall variability better than the existing crops. While selecting alternate crops, thought has to be given about the availability of the seed, special management practices required, if any for the introduced crop and accordingly farmers are made aware of the requirements of the crop.

In Yagantipalli village of Kurnool, AP, climate resilient crops served as an insurance against crop failure due to drought. Farmers traditionally grow cotton which gives lower yields during drought years which is common in this village and farmers incur losses due to higher cost of cultivation and risky nature in shallow soils. Short duration foxtail millet varieties (SIA-3085 and Suryanandi) were introduced which were found to be remunerative, tolerant to drought and can perform better than the existing crop. It was introduced in 2012 in 12 ha area and the average yields were 19.5q/ha with net returns of Rs 11,820/ha apart from the biomass used as fodder. The short duration varieties matured early by 10-15 days compared to local check and reduced the adverse impact of terminal drought condition.

Planting Methods

Planting methods either the ridge and furrow sowing, or the bed and furrow sowing or the broad bed and furrow sowing provides opportunities for moisture storage and also provides opportunities for draining the excess water in the event of heavy storms in black soils thus reducing the impact of both the drought as well as intense storms during the cropping season. It prolongs the availability of moisture during the drought and reduces the stagnation of water during heavy storms thus reducing the adverse impacts on crop and contributes to enhancement of productivity and minimizes the impacts of rainfall variability. It also provides better environment for seed germination and crop growth particularly in black soils.

11.12   Strategies Based on Resource Conservation and Management

Though these technologies are not crop specific but when taken up in combination with crops can contribute significantly towards enhancing the productivity and resilience. There are large number of soil, water and nutrient management technologies which contribute to both adaptation and mitigation. Much of the research done in rainfed agriculture in India relates to conservation of soil and rain water and drought proofing which is an ideal strategy for adaptation to climate change. Important technologies include *in situ* moisture conservation, rainwater harvesting and recycling, efficient use of irrigation water, conservation agriculture, energy efficiency in agriculture and use of poor quality water. Watershed management is now considered an accepted strategy for development of rainfed agriculture. Watershed approach has many elements which help both in adaptation and mitigation. For example, soil and water conservation works, farm ponds, check dams etc., moderate the runoff and minimize floods during high intensity rainfall. The plantation of multi-purpose trees in degraded lands helps in carbon sequestration. The crop and soil management practices can be tailored for both adaptation and mitigation at the landscape level.

11.13   Rainwater Conservation and Harvesting

These are based on in situ and ex-situ conservation of rainwater for recycling to rainfed crops. Selection of appropriate measures of soil and water conservation helps in harvesting as much water as possible and prolongs the availability of moisture thus enables the crop to withstand moisture stress for long period. The arresting of soil loss contributes to reduced carbon losses. Lal (2004) estimated that if water and wind erosion are arrested, it can contribute 3 to 4.6 Tg year^{-1} of carbon in India. Increased ground water utilization and pumping water from deep tube wells contributes to GHG emissions in agriculture. If surface storage of rainwater in dug out ponds is encouraged and low lift pumps are used to lift that water for supplemental irrigation, it can reduce dependence on ground water. Sharma et al estimated that about 28 m ha of rainfed area in eastern and central states has the maximum potential to generate runoff of 114 billion cubic meters which can be used to provide one supplemental irrigation in about 25 m ha of rainfed area. For storing such quantum of rainwater about 50 million farm ponds are required. This is one of the most important strategies not only to control runoff and soil loss but also contribute to climate change adaptation and mitigation.

In Nacharam village of Khammam, farm ponds brought about a perceptible change in crop production during *Kharif* season. Though the rainfall was less (6 rainy days) during months of June and early part of July, the intense storms of 39.3 mm and 34.8 mm generated runoff and was stored in farm ponds created in farmers field. The harvested water was used for critical irrigations to cotton, chilies and fodder grass. Farmers realized an additional yield and income from cotton (250 kg/ha, Rs.10500), chilies (150 kg/ha, Rs.9000) and fodder grass (4 t/ha, Rs.10900).

11.14   Agroforestry

Agroforestry systems like agri-silvi-culture, silvipasture and agri-horticulture offer both adaptation and mitigation opportunities.

Agroforestry systems buffer farmers against climate variability, and reduce atmospheric loads of greenhouse gases. Agroforestry can both sequester carbon and produce a range of economic, environmental, and socioeconomic benefits; the extent of sequestration can be upto 10 t ha^{-1} year^{-1} in short rotation Eucalyptus, leucaena plantations (Table 11.4). Agrisilviculture systems with moderate tree density with intercrops have however lower potential.

Table 11.4 Carbon storage (Mg/ha/ year) in different Agri silvicultural systems

Location	System	Carbon sequestration (Mg ha^{-1} year^{-1})	Reference
Raipur	Gmelina based system	2.96*	Swami & Puri, 2005
Chandigarh	Leucaena based system	0.87	Mittal & Singh, 1989
Jhansi	Anogeissus based system	1.36	Rai et al 2002
Coimbatore	Casuarina based system	1.45	Viswanath et al. 2004

*Includes soil carbon storage of 0.42 Mg ha^{-1} year^{-1} (upto 60 cm depth)

Conclusions

Climate resilient practices for a crop or cropping systems are location specific and driven by the management practices and the governed by the nature of production systems itself. The rainfed production systems have altogether different kind of resilient practices compared to an irrigated system. In case of rainfed system, the resilient practices aim at addressing the climatic vulnerability particularly the rainfall variability and measures aiming at preventing the occurrence of droughts during the cropping season and minimizing the impacts of droughts on the crop and stabilizing income during the droughts by way of other components of the farming system. In case of irrigated systems, the technologies aim at resource use efficiency, minimizing the emissions and addressing some of the issues like residue burning, etc. Selection of appropriate resilient technology for a system depends on the above aspects and to be deployed carefully depending on the climate exigencies.

The variability in rainfall being witnessed in the rainfed regions of Karnataka, Andhra Pradesh and Telangana had significant impact on the crop growth and production. Crop cultivars which can tolerate the stresses coupled with efficient natural resource management plays an important role in reducing the adverse due to these events. Demonstration of technologies in various vulnerable regions of the country is an important step in enhancing the awareness and demonstrating the effectiveness of the technologies and also contributes towards the scaling up of these technologies. There is a need to sensitize the policy makers about the worth of these technologies so as to make them part of the policy for reaching large sections of the farming community for enhancing their adaptive capacity.

References

Agarwal, P.K. (2009). Global Climate Change and Indian Agriculture; case Studies from ICAR Network Project, Indian Council of Agricultural Research, 148p.

IPCC, (2013). Summary for Policymakers. In: Climate Change 2013: The Physical Science Basis. Contribution of Working Group I to the Fifth Assessment Report of the Intergovernmental Panel on Climate Change [Stocker, T.F., Qin, .D., Plattner, G.K., Tignor, M., Allen, S.K., Boschung, J., Nauels, A., Xia, Y., Bex, V., Midgley, P.M., (eds.)]. Cambridge University Press, Cambridge, United Kingdom and New York, NY, USA.

Krishna Kumar. (2009). Impact of Climate Change on India's Monsoon Climate and Development of High Resolution Climate Change Scenarios for India. Presented at MoEF, New Delhi on October 14, 2009 (http: moef.nic.in).

Lal, R. (2001). Future Climate Change: Implications for India Summer Monsoon and its Variability, Current Science 81: 1205-1207.

Lal, R. (2004). Soil Carbon Sequestration in India. Climatic Change 65: 277–296p.

Maheswari M, Sarkar B, Vanaja M, Srinivasa Rao M, Srinivasa Rao Ch, Venkateswarlu B, Sikka AK (2015). Climate Resilient Crop Varieties for Sustainable Food Production under Aberrant Weather Conditions. CVRIDA, Hyderabad; 47p.

Mall, R., Singh, Ranjeet; Gupta, Akhilesh; Srinivasan, G.; Rathore, L. (2006). Impact of Climate Change on Indian Agriculture: A Review. Climatic Change, 78: 445-478p.

Mittal, S.P. and Singh, P. (1989). Intercropping Field Crops between Rows of *Leucaena Leucocephala* under Rainfed Conditions In Northern India. *Agroforestry Systems*, 8: 165-172p.

Network Project on Climate Change (NPCC) Consolidated Annual Report (2004-2007). Central Research Institute for Dryland Agriculture, Hyderabad, 321p.

Network Project on Climate Change (NPCC) Annual Report (2009-2010). Central Research Institute for Dryland Agriculture, Hyderabad, 296p.

National Initiative on Climate Resilient Agriculture (NICRA) (2013). AICRIPAM Component: Annual Report-2013. Central Research Institute for Dryland Agriculture, Hyderabad, India.

Prasad, YG., Srinivasa Rao, Ch., Prasad, JVNS., Rao, KV., Ramana, DBV., Gopinath, KA., Srinivas, I., Reddy, BS., Adake, R., Rao, VUM., Maheswari, M., Singh, AK and Sikka, AK. (2015). Technology Demonstrations: Enhancing Resilience and Adaptive Capacity of Farmers to Climate Variability. National Innovations in Climate Resilient Agriculture (NICRA) Project, ICAR-CRIDA, Hyderabad. 109p.

Rai, A.K., Solanki, K.R. and Rai, P. (2002). Performance of *Anogeissus Pendula* Genotypes under Agrisilvicultural system. *Indian Journal Agroforestry*, 4(1): 71 -74p.

Saseendran, A.S.K., Singh, K.K., Rathore, L.S., Singh, S.V. and Sinha, S.K., (2000). Effects of Climate Change on Rice Production in the Tropical Humid Climate of Kerala, India, Climate Change, 44: 495-514p.

Swamy, S. L. and Puri S. (2005). Biomass Production and C-sequestration of *Gmelina arborea* in Plantation and Agroforestry System in India. *Agroforestry systems* 64(3): 181-195p.

Viswanath, S., Peddappaiah, R.S., Subramoniam, V., Manivachakam, P. and George, M. (2004). Management of *Casuarina equisetifolia* in Wide-row Intercropping Systems for Enhanced Productivity. *Indian Journal of Agroforestry* 6(2): 19-25p.

12

Farmer's Knowledge, Attitude, Sources of Information and Adaption Measures towards Climate Change

K. Ravi Shankar, K. Nagasree, G. Nirmala, M.S. Prasad and Ch. Srinivasa Rao

Central Research Institute for Dryland Agriculture (CRIDA), Hyderabad.

12.1 Introduction

Indigenous knowledge is generally defined as the "Knowledge of a people of a particular area based on their interactions and experiences within that area, their traditions, and their incorporation of knowledge emanating from elsewhere into their production and economic systems". It is culturally appropriate, holistic and integrative. Incorporating indigenous knowledge is less expensive than bringing in aid for populations unprepared for catastrophes and disasters, or than importing adaptive measures, which are usually introduced in a top- down manner and difficult to implement, particularly because of financial and institutional constraints. Indigenous people that live close to natural resources often observe the activities around them and are the first to identify and adapt to any changes. The appearance of certain birds, mating of certain animals and flowering of certain plants are all important signals of changes in time and seasons that are well understood in traditional knowledge systems. Farmers have been confronted with changing environments for millennia and have developed a wide array of coping strategies, and their traditional knowledge and practices provide an important basis for facing the even greater challenges of climate change. The increasing attention to adaptation to climate change has not come with sufficient emphasis on the local nature of climate adaptation and on the role of local institutions and local governance in shaping adaptation practices (Agrawal *et al.*, 2009). Local knowledge is therefore a major priority in the planning of adaptation (Allen, 2006). Indigenous knowledge systems can facilitate understanding and effective communication and increase the rate of dissemination and utilization of climate change mitigation and adaptation options. While the importance of indigenous knowledge has been realized in the design and implementation of sustainable development projects, little attention has been drawn to their incorporation into formal changing climate mitigation and adaptation strategies (Nyong *et al.*, 2007).

12.2 Climate Change: Impacts on Agriculture in India

Agriculture is one of the largest contributors to India's Gross Domestic Product (GDP), approximately 20%. It is the main source of livelihood for almost 60% of the country's total population. The impacts of changing climate on agriculture will therefore be severely felt in India. It has been projected that under the scenario of a 2.5 ºC to 4.9 ºC temperature rise in India, rice yields will drop by 32%-40% and wheat yields by 41-52%. This would cause GDP to fall by 1.8%-3.4% (GOI, 2011; Guiteras, 2007; OECD, 2002).

12.3 Adaptation to Climate Change

Agriculture in developing countries is one of the most vulnerable sectors of the global economy to changing climate (Kurukulasuriya *et al.*, 2006; Seo and Mendelsohn 2008c). Farmers whose livelihoods depend on the use of natural resources are likely to bear the brunt of adverse changing climate impacts. Farmers will be hard hit if they do not adjust at all to new climates (Mendelsohn *et al.*, 1994, Rosenzweig and Hillel 1998; Reilly *et al.*, 1996). Adaptation to climate change requires that farmers first notice that climate has changed, and then identify useful adaptations and implement them (Maddison, 2006). Adaptation is widely recognized as a vital component of any policy response to climate change. Studies show that without adaptation, climate change is generally detrimental to the agriculture sector; but with adaptation, vulnerability can largely be reduced (Easterling *et al.*, 1993; Rosenzweig and Parry 1994; Smith 1996; Mendelsohn 1998;

Reilly and Schimmelpfennig 1999; Smit and Skinner, 2002). The degree to which an agricultural system is affected by climate change depends on its adaptive capacity. The adaptive capacity of a system describes its ability to modify its characteristics or behavior so as to cope better with changes in external conditions. Adaptive capacity is determined by various factors including recognition of the need to adapt, willingness to undertake adaptation, and the availability of, and ability to deploy, resources (Brown, 2010). Recent empirical studies indicate that farmers have already adapted to the existing climates that they face by choosing crops or livestock or irrigation (Kurukulasuriya and Mendelsohn 2007, 2008; Nhemachena and Hassan 2007; Seo and Mendelsohn 2008a, 2008b) ideal for their current climate. The adaptation strategies must not be used in isolation. For instance, the use of early maturing crop varieties must be accompanied by other crop management practices such as crop rotation or the use of cover crops. This, however, requires additional institutional support, such as credit, access to input and output markets and information.

The objectives of the present study was to identify farmers' knowledge, attitude, sources of information towards climate change along with their farm-level adaptation measures with a view to suggest appropriate research/policy issues which help in facilitating farmers' adaptation to changing climate.

12.4 Research Results

A sample of 180 farmers @ 60 each from Anantapur, East Godavari districts of A.P. and Mahbubnagar district of Telangana was selected randomly. Data was collected using a pre-tested interview schedule from the farmers. Percent analysis, correlation and regression, and composite index developed in the study were used for analyzing data.

(a) *Farmers' perceptions and Adaptation Measures towards Climate change:* From table 12.1, it is clear that rise in temperatures followed by decrease in rainfall, advanced onset of monsoon, middle long dry spells, terminal heavy rains, prevalence of pests and diseases and ITKs for weather forecast failing are the major farmers' perceptions in that order of magnitude regarding climate change in Anantapur. Bryan et al. (2009) in their study in Ethiopia and South Africa reported that farmers experienced increased temperature and decreased rainfall. Similar observations were reported by Vedwan and Rhoades (2001), Hageback et al. (2005), Maddison (2006), Gbetibouo (2009) and Dejene (2011) in their studies. Results of a study conducted in Bundi district of Rajasthan, India revealed farmers' perceptions to climate change as increase in temperatures, decreased rainfall and long dry spells. The chief adaptation measures followed by farmers' are change in planting time, intercropping, soil and water conservation and planting drought tolerant crops (Dhaka *et al.*, 2010).

Table 12.1 Farmers' Perceptions regarding Climate change in Anantapur

S.No.	Farmers' Perception	Number*	%	Rank
1.	Rise in temperatures	57	95	I
2.	Decrease in rainfall	56	93	II
3.	Advanced onset of monsoon	54	90	III
4.	Middle, long dry spells	53	88	IV
5.	Terminal heavy rains	50	83	V
6.	Uneven distribution of rainfall thereby, affecting length of growing season	49	82	VI
7.	Prevalence of pests and diseases	47	78	VII
8.	ITKs for weather forecast failing	41	68	VIII

***Multiple responses**

Table 12.2 Farmers' Adaptation Measures towards Climate change in Anantapur

S.No.	Farmers' Adaptation Measures	Number*	%	Rank
1.	Buy insurance	56	93	I
2.	Change in planting dates of groundnut (go for early sowings may be between May end to early June)	55	92	II
3.	Intercrop with red gram in 8:1 or 12:1 ratio.	48	80	III
4.	Intercrop with castor contemplated	47	78	IV
5.	Construct water harvesting structures under MGNREGA	45	75	V
6.	Require quick maturing, drought resistant varieties	42	70	VI

It is clear from Table 12.2, that buying insurance, changing planting dates of groundnut, intercrop with red gram, construct water harvesting structures, and require quick maturing, drought resistant varieties in that order of magnitude are the major adaptation measures followed by farmers' towards climate change in Anantapur. This finding is consistent with that of Swanson *et al.*, (2008) who reported that crop insurance was widely used by farmers in Foremost and Coaldale regions of Canada and the common feeling was that even though it might not provide sufficient returns for losses incurred it does offer some protection. It has allowed them to continue farming. Agricultural insurance can help people to cope with the financial losses incurred as a result of weather extremes. Insurance supports farmers in their adaptation process and prevents them from falling into absolute poverty. Apart from stabilizing household incomes by reducing the economic risk, insurance can also enhance farmers' willingness to adapt, to make use of innovations and invest in new technologies (Ilona *et al.*, 2011). Agricultural adaptation involves two types of modifications in production systems. The first is increased diversification that involves engaging in production activities that are drought tolerant and or resistant to temperature stresses as well as activities that make efficient use and take full advantage of the prevailing water and temperature conditions, among other factors. Crop diversification can serve as insurance against rainfall variability as different crops are affected differently by climate events (Orindi and Eriksen 2005; Adger *et al.*, 2003). The second strategy focuses on crop management practices geared towards ensuring that critical crop growth stages do not coincide with very harsh climatic conditions such as mid-season droughts. Crop management practices that can be used include modifying the length of the growing period and changing planting and harvesting dates (Orindi and Eriksen, 2005).

From Table 12.3, it is clear that rise in temperatures followed by decrease in rainfall, prolonged dry spells in between rains, terminal heavy rains and prevalence of pests and diseases (powdery mildew, mold in castor; smut and jassids in paddy) are the major farmers' perceptions in that order of magnitude regarding climate change in Mahbubnagar. It is striking that farmers across the world show a remarkable unanimity in observations of seasonal change, particularly regarding rain falling in most intense bursts; and generally higher temperatures and longer hot, dry spells within rainy seasons, with effects on soil moisture (Jennings and Magrath, 2009). Kemausuor *et al.*, (2011) reported that a large percentage (93%) of farmers was of the opinion that the timing of the rains is now irregular and unpredictable.

Table 12.3 Farmers' Perceptions regarding Climate change in Mahbubnagar

S.No.	Farmers' Perception	Number*	%	Rank
1.	Rise in temperatures	55	92	I
2.	Decrease in rainfall	53	88	II
3.	Advanced (some places timely) onset of monsoon	51	85	III
4.	Middle long dry spells accompanied by cloudy weather during flowering	48	80	IV
5.	Terminal heavy rains	46	77	V
6.	Prevalence of pests and diseases (powdery mildew, mold in castor; smut and jassids in paddy)	41	68	VI

It is clear from Table 12.4, that staggered sowings, change in planting dates, require drought resistant crops, and construct water harvesting structures are the major adaptation measures followed by farmers' towards changing climate in Mahbubnagar.

Table 12.4 Farmers' Adaptation Measures towards Climate change in Mahbubnagar

S.No.	Farmers' Adaptation Measures	Number*	%	Rank
1.	Staggered sowings (dry paddy, castor, red gram and cotton in kharif), (groundnut, paddy, chillies and tobacco in rabi)	50	83	I
2.	Change in Planting dates and planting different crops	49	82	II
3.	Require drought resistant varieties	45	75	III
4.	Water harvesting structures started under MGNREGA	41	68	IV

Also, the farmers' in Mahbubnagar are accustomed to observe the rainy season and if the season is favourable with good rains, they will continue farming. Otherwise, they migrate and work as construction labour at Gangavati, Hyderabad and Bangalore. Higher temperatures, pest and disease attack on crops were the chief perceptions of farmers towards climate change, while, planting different crops and water conservation were the main adaptation strategies of farmers in Ogbomosho agricultural zone of Oyo state in Nigeria. (Ayanwuyi *et al.*, 2010).

From Table 12.5, it is clear that rise in temperatures, followed by decrease in rainfall, incidence of pests and diseases, terminal heavy cyclonic rains, and ITKs for rain forecasts failing are the major farmers' perceptions in that order of magnitude regarding climate change in East Godavari.

Table 12.5 Farmers' Perceptions regarding Climate change in East Godavari

S.No.	Farmers' Perception	Number*	%	Rank
1.	Rise in temperatures.	54	90	I
2.	Decrease in rainfall.	53	88	II
3.	Pest and disease incidence is high for kharif paddy like BPH, BLB and stem borer (at transplanting stage).	51	85	III
4.	Terminal heavy and unseasonal rains.	49	82	IV
5.	ITKs for rain forecasts are failing.	45	75	V

It is clear from Table 12.6, that early sowings, salt water spray for harvested paddy stalks, strengthening of river banks and improved drainage, survey number wise insurance, and loans to tenant farmers are the major adaptation measures perceived by farmers' towards climate change in East Godavari. Migrate as construction labour if monsoon fails, particularly in rainfed areas of the district is another common phenomenon (Ravi Shankar *et al.*, 2013). Improving the adaptive capacity of disadvantaged communities requires ensuring access to resources, income generation activities, greater equity between genders and social groups, and an increase in the capacity of the poor to participate in local politics and actions (IISD 2006).

Table 12.6 Farmers' Adaptation Measures towards
Cimate change in East Godavari

S.No.	Farmers' Adaptation Measures	Number*	%	Rank
1.	Go for early (June) sowings to avoid November cyclones coinciding with harvests.	56	93	I
2.	Salt water spray for harvested paddy stalks to avoid discoloration and regermination. For paddy in field, tying with rope and sticks on four sides to keep them erect and not falling down.	55	92	II
3.	Strengthening of river banks and improved drainage.	53	88	III
4.	Survey number wise insurance covering low lands.	50	83	IV
5.	Loans to tenant farmers though introduced, falls short of actual requirements in terms of coverage.	48	80	V

Since most smallholder farmers are operating under resource limitations, lack of credit facilities and other inputs compound the limitations of resource availability and the implications are that farmers fail to meet transaction costs necessary to acquire the adaptation measures they might want to and at times farmers cannot make beneficial use of the available information they might have (Kandlinkar and Risbey 2000). Lack of access to credit has been observed in previous studies (Nhemachena and Hassan, 2007) to be a barrier to responding to climate change. A better understanding of how farmers' perceive changing climate, ongoing adaptation measures, and the factors influencing the decision to adapt farming practices is needed to craft policies and programmes aimed at promoting successful adaptation of the agricultural sector (Bryan *et al.*, 2009).

- The mean adaptation index value for floods (12.13) (East Godavari) is greater than that for droughts (11.90, 11.65) (Anantapur and Mahbubnagar respectively).

- Practices like construct water harvesting structures, plant drought resistant crops, crop management by adjusting planting dates and soil management by mulching, conservation tillage showed highest adaptation in Anantapur and Mahbubnagar. This amply illustrates the need for water harvesting, storage and reuse.

- Practices like flood forecasting and early warning systems, drainage aspects, better soil and crop management and community based water management showed highest adaptation in East Godavari. The problem here is managing excess water.

- Education, farming experience, and farm size were contributing significantly at 0.01 level with farmers' adaptation to climate change.

Results from Akola, Solapur (Maharashtra) and Bijapur (Karnataka)

A study was conducted in Maharashtra and Karnataka where All India Coordinated Research Project on Dryland Agriculture (AICRPDA) centers (3) are located duly reflecting drought conditions in red and black soils. The results are based on sample size of 180 farmers that include 30 farmers from each village and two villages from each centre.

In Akola, 92% of farmers **perceived** that temperatures have increased; 63% expressed that rains are delayed and have become shorter; 45% expressed prolonged dry spells; and 43% of them expressed extended breaks in monsoon. The chief **adaptation** measures followed by farmers towards climate variability and change are: 68% of them expressed that they are changing planting dates and cropping pattern; 60% of them depend on timely availability of inputs; 60% of them go for drought resistant crops; 53% depend on contingency crop planning and 52% of them spray urea.

In Solapur, the chief farmers' **perceptions** are: 63% of farmers expressed prolonged dry spells; 48% expressed that rains are delayed and have become shorter; 42% of them expressed changes in onset of monsoon and extended monsoon periods and 32% expressed extended breaks in monsoon. The chief **adaptation** measures followed by farmers towards climate variability and change are: 77% of them expressed that they are changing planting dates and cropping pattern; 65% are planting different crops; 50% are construct water harvesting structures; 27% diversify to livestock or other agricultural enterprises and 15% of them are buying insurance.

In Bijapur, chief farmers' **perceptions** are extended monsoon periods expressed by 52% of them; 50% expressed that rains are delayed and have become shorter; 28% of them each expressed extended breaks in monsoon and rise in temperatures; and 27% of them reported prolonged dry spells. The chief **adaptation** measures towards climate variability and change are: 45% of them expressed that they are changing planting dates and cropping pattern; 35% of them plant different crops; 30% of them spray urea; 25% of them are buying insurance and 23% of them diversify to livestock or other agricultural enterprises.

Barriers to climate change adaptation in Solapur as expressed by farmer's are: lack of access to credit (53%); lack of savings (42%); lack of labour (38%); lack of access to water (35%) and lack of market access (38%). Barriers to climate change adaptation in Bijapur as expressed by farmer's are: lack of access to credit (50%); lack of labour (40%); unremunerative prices (32%); lack of access to water (22%) and lack of information on appropriate adaptation measures (22%).

The mean adaptation index expressed in % for Akola is 38, Solapur is 32.96, and for Bijapur it is 29.07.

12.5 Different Sources of Information on Climate Change

TV, Radio, Newspapers, other farmers (especially for weather forecasts) and officials are the major sources of information available to farmers regarding climate change.

12.6 Attitude towards Climate Change

Attitude in this study means the degree of positive or negative feelings, beliefs of farmers towards climate variability and change in agriculture and allied fields. The steps followed for constructing the Likert (1932) type of scale to measure the attitude of farmers towards climate variability is discussed below:

(i)   Collection of statements.

(ii) Editing of the statements.

(iii) Selection of statements and scoring technique.

(iv) Critical ratio (t value)

(v)   Reliability.

(vi) Validity.

Attitudes of farmers' towards climate change provide feedback to the research for developing tools for the decision support systems. The attitude of farmers' towards climate change was measured by developing a 20 statement questionnaire with a five point continuum. The chief attitudes with which farmers' agreed are:

- Climate change is a serious problem.
- Climate change is affecting my farming.
- Average temperatures are increasing.
- Human activity is responsible for climate change.
- Climate change affects small and marginal farmers more.
- Climate change impacted food production of my farm.
- Cropping seasons in my village are changing.
- Rainfall patterns are changing.
- Climate change enhanced incidence of pests and diseases.
- Government should do more to help farmers adapt to climate change.
- Scientists/Govt. will solve the problems of climate change.

Conclusion

Capacity building at local, national and regional levels is vital to enable developing countries like India to adapt to climate change. Capacity building, for example atple to integrate climate change and socio-economic assessments into vulnerability and adaptation assessments, helps to better identify effective adaptation options and their associated costs. Education and training of stakeholders, including policy-level decision makers, are important catalysts for the success of assessing vulnerabilities and planning adaptation activities, as well as implementing adaptation plans. Extension can make a significant contribution through enhanced farmer decision making in the light of climate change. The most important purpose for extension today is to bring about the empowerment of farmers, so that their voices can be heard and they can play a major role in deciding how they will mitigate and adapt to climate change.

As national and international policy makers turn their attention to climate change adaptation, they should keep in mind that constructing an enabling environment that minimizes these vulnerabilities will be central to any meaningful and lasting increase in the adaptive capacity of the rural poor. Govt. policies designed to promote adaptation at the farm level will lead to greater food and livelihood security in the face of climate change.

References

Adger W.N., S. Huq, K. Brown, D. Conway and M. Hulme. (2003). Adaptation to Climate Change in the Developing World. *Progress in Development Studies* 3: 179-195p.

Agrawal, A., Kononen, M. and Perrin N. (2009). The Role of Local Institutions in Adaptation to Climate Change. *Social Development working papers*, 118p.

Allen, K.M. (2006). Community based Disaster Preparedness and Climate Adaptation: Local Capacity Building in the Philippines. *Disasters* 30(1): 81-101p.

Ayanwuyi, E. Kuponiyi, F.A. Ogunlade and Oyetoro J.O. (2010). Farmers Perception of Impact of Climate Changes on Food Crop Production in Ogbomosho Agricultural Zone of Oyo State, Nigeria. *Global Journal of Human Social Science*, Vol.10, Issue 7: 33-39p.

Brown, K. (2010). Climate Change and Development Short Course: Resilience and Adaptive Capacity (Power Point presentation). Norwich, UK: International Development, University of East Anglia.

Bryan, E., Deressa, T.T, Gbetibouo, G.A. and Ringler, C. (2009). Adaptation to Climate Change in Ethiopia and South Africa: Options and Constraints. *Environmental Science and Policy*, 12(4): 413-426p.

Dejene K. Mengistu (2011). Farmers' perception and knowledge of climate change and their coping strategies to the related hazards: Case study from Adiha, Central Tigray, Ethiopia. *Agricultural Sciences* 2: 138-145p.

Dhaka, B.L., Chayal, K. and Poonia, M.K. (2010). Analysis of Farmers' Perception and Adaptation Strategies to Climate Change, *Libyan Agricultural Research Center Journal International*, 1 (6): 388-390p.

Easterling, W.E., P.R. Crosson, N.J. Rosenberg, M.S. McKenney, L.A. Katz, and K.M. Lemon (1993). Agricultural Impacts of and Responses to Climate Change in the Missouri-Iowa-Nebraska Region. *Climatic Change*, 24(1-2): 23-62p.

Gbetibouo, G.A. (2009). Understanding Farmers' Perceptions and Adaptations to Climate Change and Variability: The Case of the Limpopo Basin, South Africa. Discussion Paper No. 00849, South Africa Environment and Production Technology Division, IFPRI.

GOI-Government of India 2011: Agriculture. http://india.gov.in /sectors/agriculture/index.php

Guiteras, R. (2007). The Impact of Climate Change on Indian Agriculture.

http://www.colgate.edu/portaldata/imagegallerywww/2050/ImageGa llery/GuiterasPaper.pdf

Hageback, J.M. Sundbery, D. Ostroald, X. Chen, and P. Knutsson. (2005). Climate Variability and Land use Change in Danagou Watershed, China-Examples of Small Scale Farmers Adaptation. *Climate Change* 72: 189-212p.

IFPRI – International Food Policy Research Institute, 2009: Climate Change Impact on Agriculture and Costs of Adaptation. http://www.undp-adaptation.org/undpcc/files/docs/publications/ pr21.pdf

IISD (International Institute for Sustainable Development) (2006). *Understanding adaptation to climate change in developing countries.* http://www.iisd.org. Accessed November 20.

Ilona Porsche, Anna Kalisch, and Rosie Fuglien. (2011). Adaptation in Agriculture: 40-83 In: Adaptation to Climate Change with a Focus on Rural Areas and India: 230p, New Delhi.

Jennings, S. and Magrath, J., (2009). What Happened to the Seasons? Oxfam Research Report. Available online at: http://www .oxfam.org.uk/resources/policy/climate_change/research-where-are-the-seasons.html

Kandlinkar, M. and J. Risbey. (2000). Agricultural Impacts of Climate Change: If Adaptation is the Answer, What is the Question? *Climatic Change* 45: 529-539.

Kemausuor, F., Dwamena, E., Ato Bart-Plange and Nicholas Kyei Baffour. (2011). Farmers' Perception of Climate Change in the Ejura-Sekyedumase District of Ghana, *ARPN Journal of Agricultural and Biological Science*, Vol.6, No.10: 26-37.

Kurukulasuriya, P. and R. Mendelsohn. (2007). Modeling Endogenous Irrigation: The Impact of Climate Change on Farmers in Africa". World Bank Policy Research Working Paper 4278.

Kurukulasuriya, P. and R. Mendelsohn. (2008). Crop Switching as an Adaptation Strategy to Climate Change African Journal Agriculture and Resource Economics (forthcoming).

Kurukulasuriya, P., R. Mendelsohn, R. Hassan, J. Benhin, M. Diop, H. M. Eid, K.Y. Fosu, G.Gbetibouo, S. Jain, A. Mahamadou, S. El-Marsafawy, S. Ouda, M. Ouedraogo, I. S, D. Maddision, N. Seo and A. Dinar. (2006). Will African Agriculture Survive Climate Change? World Bank Economic Review 20(3), 367-388p.

Likert, R. (1932). A Technique for the Measurement of Attitudes. *Arch. Psychol.* No. 140.

Maddison, D. (2006). *The Perception of Adaptation to Climate Change in Africa.* CEEPA. Discussion Paper No. 10. Centre for Environmental Economics and Policy in Africa. Pretoria, South Africa: University of Pretoria.

Mendelsohn, R. (1998). Climate-change Damages. In *Economics and Policy Issues in Climate Change*, ed. W.D. Nordhaus. Resources for the Future: Washington, D.C.

Mendelsohn, R., W. Nordhaus and D. Shaw. (1994). The Impact of Global Warming on Agriculture: A Ricardian Analysis, *American Economic Review* 84, 753-771p.

Nhemachena, C. and R Hassan. (2007). Determinants of climate adaptation strategies of African farmers: Multinomial choice analysis, Draft Report, CEEPA, University of Pretoria.

Nyong, A., Adesina, F. and Elasha, B.O. (2007). The Value of Indigenous Knowledge in Climate Change Mitigation and Adaptation Strategies in the African Sahel. *Mitigation and Adaptation Strategies in Global Change* 12:787-797.

OECD, (2002). Organisation for Economic Cooperation and Development 2002: Climate Change: India's perceptions, positions, policies and possibilities.

http://www.oecd.org/dataoecd/22/16/1934784.pdf

Orindi, V.A., and S. Eriksen. (2005). Mainstreaming Adaptation to Climate Change in the Development Process in Uganda. *Ecopolicy Series* 15. Nairobi, Kenya: African Centre for Technology Studies (ACTS).

Ravi Shankar, K., Nagasree, K., Prasad, M.S., and Venkateswarlu, B. (2013). Farmers' Knowledge Perceptions and Adaptation Measures towards Climate Change in South India and Role of Extension in Climate Change Adaptation and Mitigation. *In:* Compendium of National Seminar on Futuristic Agricultural Extension for Livelihood Improvement and Sustainable Development, January 19-21, 2013 at ANGRAU, Rajendranagar, Hyderabad: 295-303p.

Reilly, J., and D. Schimmelpfennig. (1999). Agricultural Impact Assessment, Vulnerability and Scope for Adaptation. *Climatic Change* 43: 745-788p.

Reilly, J., W. Baethgen, F.E. Chege, S.C. van de Greijn, L. Ferda, A. Iglesia, C. Kenny, D. Patterson, J. Rogasik, R. Rotter, C. Rosenzweig, W. Sombroek, and J. Westbrook (1996). Agriculture in a changing climate: impacts and adaptations, in IPCC (Intergovernmental Panel on Climate Change), R.Watson, M. Zinyowera, R. Moss, and D. Dokken (eds), *Climate Change 1995: Impacts, Adaptations, and Mitigation of Climate Change: Scientific-Technical Analyses*, Cambridge: Cambridge University Press, Cambridge pp. 427- 468p.

Rosenzweig, C., and D. Hillel (1998). Climate Change and the Global Harvest: Potential Impacts of the Greenhouse Effect on Agriculture. Oxford University Press, New York, USA.

Rosenzweig, C, and M.L. Parry. (1994). Potential Impact of Climate-Change on World Food Supply. *Nature* 367: 133-138.

Seo, S. N. and R. Mendelsohn. (2008a). Measuring Impacts and Adaptations to Climate Change: A Structural Ricardian Model of African Livestock Management, *Agricultural Economics*, 38, 1-15p.

Seo, S. N. and R. Mendelsohn. (2008b). An Analysis of Crop Choice: Adapting to Climate Change in South American Farms, *Ecological Economics* (forthcoming).

Seo, S. N. and R. Mendelsohn. (2008c). A Ricardian Analysis of the Impact of Climate Change on South American Farms, *Agricultura Technica* (forthcoming).

Smit B., and M.W. Skinner. (2002). Adaptation Options in Agriculture to Climate Change: A Typology. *Mitigation and Adaptation strategies for Global Change* 7: 85-114p.

Smith, J.B. (1996). Using a decision Matrix to Assess Climate Change Adaptation. In *Adapting to climate change: An international Perspective*, ed. J.B. Smith, N. Bhatti, G. Menzhulin, R. Benioff, M.I. Budyko, M. Campos, B.Jallow, and F. Rijsberman. New York: Springer.

Swanson, D., Henry David, V., Christa Rust, and Jennifer Medlock (2008). Understanding Adaptive Policy Mechanisms through Farm-Level Studies of Adaptation to Weather Events in Alberta, Canada. Published by the International Institute for Sustainable Development, Canada: 72p.

UNFCC – United Nations Framework Convention on Climate Change 2009: Climate Change Impacts, vulnerabilities and adaptation in developing countries. http://unfcc.int/resource/docs/publications /impacts.pdf

Vedwan, N. and R.E. Rhoades. (2001). Climate Change in the Western Himalayas of India: A Study of Local Perception and Response. *Climate Research* 19: 109-117p.

13

Community Based Approaches to Climate Smart Agriculture

Rajeswar Jonnalagadda

Interco-operation Social Development India (ISDI), Hyderabad.

13.1 Climate Change and Agriculture

The contribution of agriculture and allied sectors to national GDP, which was 51.9 per cent of GDP in 1950-51 has been declining at an alarming rate, due to shift from traditional agrarian economy to industry and service sectors. The whole of agriculturalsectorthat engages 49 percent of the workforce of the country has contributeda mere 13.9 percent of Gross Domestic Product (GDP)of the country during 2013-14, according to theCentral Statistics Officeand Economic Survey 2015, Government of India.With two successive years of almost drought like scenario in 2015, with monsoon rains being deficient compared to long period average, there has been a dismal growth rate of 1.1 percent, against a target of at least 4 percentgrowth rate in agriculture sector. If agriculture has to grow at 4 percent as targeted for the 12thPlan as a whole, it must attain 7 percent growth per annum for the remaining period. However, as stated in the Fifth Assessment Report (AR5) of the Inter-Governmental Panel on Climate Change (IPCC), climate change will increase the risk of food insecurity and the breakdown of food systems through drought, flooding, and rain variability and extremes– particularly for poorer populations. The AR5 of IPCC also cautioned that by 2050, risk of hunger could increase by 20 percent and child malnutrition will be 20 percent higher than it would have been without climate change. The small and marginal farmers have to fight against all odds – whether it is increasing input-costs, unpredictable weather conditions, poor harvests and falling prices for their produce.

The ambitious National Mission for Sustainable Agriculture (NMSA) is one of the eight missions launched in the year 2010 under the National Action Plan on Climate Change (NAPCC), which was aimed at transforming Indian agriculture into a climate-resilient production system through suitable adaptation and mitigation measures in the domain of crops and animal husbandry. However, the paucity of funds for NMSA led to some goals of the mission being embedded into five existing programmes of the Agriculture and Cooperation departments – RashtriyaKrishiVikasYojana (RKVY); National Food Security Mission (NFSM); National Horticulture Mission (NHM); GraminBhandaranYojana; Integrated Scheme of Oilseeds, Pulses, Oil palm, and Maize (ISOPOM).

NMSA has been conceptualized by subsuming *Rain-fed Area Development programme* (RADP), *National Mission on Micro Irrigation* (NMMI), *National Project on Organic Farming* (NPOF), *National Project on Management of Soil Health & Fertility* (NPMSH&F) and *Soil and Land Use Survey of India,* under its domain. NMSA accords special focus for development of rain-fed areas, resource conservation, water use efficiency and soil health management. NMSA also replicates the learning of the *National Initiatives of Climate Resilient Agriculture* (NICRA) being implemented by ICAR.

Given the severity of climate change impacts, all attention should be focused on providing tools and financing for countries and communities to protect their food security and food sovereignty from the impacts of climate change.

13.2 The Carbon Economy of Agriculture

The FAO,World Bank, the Consultative Group on International Agricultural Research (CGIAR) and the Global Alliance for Climate-Smart Agriculture (GACSA) have been promoting Climate-Smart Agriculture (CSA) as a "triple-win", which enhances (i) agricultural productivity, (ii) climate-resilience through adaptation and mitigation measures and (iii) food security and ensured household income for smallholder farmers. CSA is a holistic approach that requires multi-pronged investment and a multi-disciplinary approach towards participatory research. Water is the key to enhancing resilience of production systems to climate variability and climate change. Hence, public investment in water, especially low-cost solutions which could be taken up by smallholder farmers, lies at the core of CSA. While the role of private sector investments need to be emphasised, initiatives on CSA for poverty alleviation and value addition to rural livelihoods have to be scaled up at national and regional levels. Thus, while more investments are needed especially in water, dryland agriculture, etc., more than 80 percentof investments in agricultureare coming from the private sector mostly in profitable sectors of agriculture only.

However, in the whole gamut of climate-smart agriculture, emphasis is on enhanced crop-productivity,the mitigation of greenhouse gases (GHGs) from agriculture and soil carbon sequestration.It is worth mentioning in this context that emissions of GHGs like N_2O (which has 300 times more heat-trapping potential than CO_2) are very high in developed countries where agriculture is industrialized with heavy inputs of inorganic fertilizers. The scope for mitigation of N_2O emissions from soil is marginal in poor and developing countries, when compared with industrialized countries. Moreover, commoditization of soil-carbon through market instruments like Clean Development Mechanism (CDM)as the main instrument to combat climate change may not ensure food security; but it would certainly hamper the food sovereignty of nations on one hand and would also shift the focus of climate change adaptation away from the livelihood needs and value addition to the household incomes of small and marginal farmers.

Despite global hype onCarbon Trading and Carbon-Credits, the CDM regime contributed only about 14 percent oftotal mitigation of GHGemissions till date. The success of CDM in land-use sector involving agriculture and afforestation is even more limited. The course of Carbon –Economy may yield partial results in mitigation of GHG emissionsthrough some market incentives, only in a select sectors like power, industries, infrastructure etc., which are mostly controlled by private entrepreneurs. But in a complex sector like Agriculture, Forestry and Other Land Use (AFOLU), besides carbon economy, issues like livelihoods, security and sovereignty have to be addressed through a bottom-up approach.

Need to Diversify Climate-Smart Agriculture: Enhancing the resilience of agriculture to climate change through adaptation strategies that promote the development of sustainable agriculture is a common multiple-benefit recommendation for agricultural adaptation. The Climate-Smart Agriculture should be made more inclusive to:

- Promote a research and knowledge-sharing agenda on ecological agriculture, adaptation, and climate resilience with emphasis on women farmers, and farmer, indigenous, and traditional knowledge-sharing systems

- Focus on building resilience with strong social protection programmes, such as the Food Security Climate Resilience Facility of the World Food Programme (WFP)

- Increase national and international investment in climate-resilient, ecological agriculture, including agro-ecological approaches, such as through the Adaptation for Smallholder Agriculture Programme of the International Fund for Agricultural Development (IFAD)

IFAD for Community Based Agriculture: As emphasized by the International Fund for Agricultural Development (IFAD), a specialized agency of the United Nations, the causes of food insecurity were not so much failures in food production but structural problems relating to poverty of small and marginal farmers in poor and developing countries. CSA plans integrate a host of approaches, including maximizing the use of natural processes and ecosystems, reducing external inorganic inputs and waste, diversifying production, ensuring crops and livestock are cultivated in locally appropriate proportions, and mixing new technologies with traditional knowledge (Grainger-Jones, IFAD, 2012).

The real solutions to climate change in farmers' fields must be the climate-resilient practices of ecologically based agricultural systems, not the market-oriented agricultural mitigation methods. The following adaptation measures may increase the capacity of the agricultural system to minimize the effects of climate change on productivity:

- Alternative cultivation Practices as an Adaptation measure
- Integrating livestock with crop production systems
- Improving soil quality through soil-carbon sequestration
- Enhanced Water Use Efficiency in agriculture
- Minimizing off-farm flow of nutrients and pesticides, and
- Integrating the traditional/indigenous practices for sustainable agriculture.

13.3 Adaptation for Smallholder Agriculture Programme (ASAP)

Launched by IFAD in 2012, the *Adaptation for Smallholder Agriculture Programme* (ASAP) aims at a major scaling-up of successful "multiple-benefit" approaches to increase agricultural output, while simultaneously reducing vulnerability to climate-related risks and diversifying livelihoods. ASAP facilitates climate and environmental finance to smallholder farmers so they can access the information tools and technologies that help build their resilience to climate change. ASAP has become the largest global financing source dedicated to supporting the adaptation of poor smallholder farmers to climate change.

The objective of ASAP is to improve the climate resilience of large-scale rural development programmes and improve the capacity of at least 8 million smallholder farmers to expand their options in a rapidly changing environment. Through ASAP, IFAD is driving a major scaling-up of successful "multiple-benefit" approaches to increase agricultural output while simultaneously reducing vulnerability to climate-related risks and diversifying livelihoods.

ASAP-supports initiatives that include the following, among others:

- Mixed crop and livestock systems which integrate the use of drought-tolerant crops and manure, which can help increase agricultural productivity while at the same time diversifying risks across different products.

- Systems of crop rotation which consider both food and fodder crops, which can reduce exposure to climate threats while also improving family nutrition.
- A combination of agroforestry systems and communal ponds, which can improve the quality of soils, increase the availability of water during dry periods, and provide additional income.

ASAP will empower community-based organisations to make use of new climate risk management skills, information and technologies, through:

- Improved weather stations networks, for more reliable seasonal forecasts and cropping calendars
- Geographic Information Systems for monitoring landscape use in a changing environment
- Economic valuation of climate change impacts for more robust policy decisions.
- Blending the tried and tested 'no regrets' approaches with modern adaptation know-how

Benefits of Climate Smart Agriculture (CSA): The concept of climate-smart agriculture (CSA) brings in a paradigm-shift towards sustainable consumption and production (SCP) practices in agriculture. Climate-smart agriculture is an approach that aims to integrate social, economic and ecologicalobjectives to increase agricultural yields, boost profits, reduce local pollution, address poverty, enhanceclimate resilience and reduce greenhouse gas emissions. CSA differs from other clean technology sectors because it tends to rely on donor and governmentinvolvement and depends heavily on behaviour change, education, and institutional reform, butrepresents a commercial opportunity for SMEs as well (World Bank, 2014). It pursues the triple bottom line of environment, economy and equity by implementing the following practices:

- Micro-watershed management involving local communities

- Augmenting the energy security by installing solar powered water pump sets

- Protection and conservation of topsoil of agricultural lands through sustainable production practices

- Mitigation of GHG emissions from crop production

- Integrating agriculture and livestock management for ensured household incomes to smallholder farmers

- Lifecycle approach to agriculture, involving conservation tillage, manure management, soil carbon sequestration, soil moisture retention, minimizing the external input of fertilizers, enhanced community participation involving women, introducing effective early-warning tools, establishing forward and backward linkages for agricultural produce etc.

By introducing the concept of enhanced water use efficiency, and innovative technologies like solar-powered water pump sets, CSA attempts to decouple agricultural productivity and economic growth from associated environmental degradation and greenhouse gas emissions that contribute to climate change. In nutshell, it is all about *producing more and better, with less inputs of resources*, by pursuing the following activities:

- Introducing shift towards Sustainable Consumption and Production (SCP) in agriculture sector

- Supporting the regional and national policy initiatives like the *National Mission for Sustainable Agriculture* (NMSA), the *National Initiatives of Climate Resilient Agriculture* (NICRA) etc.

- Reducing the Carbon footprint of agriculture

- Bringing in economic benefits to smallholder farmers

- Bringing social equity by encouraging women and community participation in integrated farming systems for climate-smart agriculture

- Supporting capacity building and facilitating access to financial and technical assistance

Enhanced productivity in agriculture is possible through Sustainable Consumption and Production (SCP) model with an integrated approach of conservation tillage, soil carbon sequestration, soil moisture retention, livestock feed and manure management, micro-watershed management etc. Zero-tillage or controlled tillage in climate smart agriculture not only helps in soil-carbon sequestration, but also has several benefits like prevention of soil erosion, retention of soil moisture, reducing the costs of fuel and labour associated with conventional tillage (Jat et al. 2014)

In agriculture sector, there has been a paradigm-shift from mere climate-resilience to being climate-smart. Besides adaptation to climate-induced weather extremes, mitigation of GHGs from agriculture practices is also being given equal emphasis under Climate Smart Agriculture. While the contribution of Agriculture sector to total GHG emissions is 24 percent at global level, in India it is about 17.6 percent of total GHG emissions including Carbon Dioxide, Methane and Nitrous Oxide. The emission from Indian agriculture primarily emanates from five principal activities, *viz.* enteric fermentation from livestock, rice cultivation, manure management, agricultural soils and burning of crop residues (NAAS, 2014).Indian agriculture produces about 500-550 million tons of crop residues annually. However, a large portion of this, about 90-140 million tons annually is burnton-farm primarily to clear the field for sowing of the succeeding crop (NAAS, 2012).

However, by synergizing traditional indigenous knowledge with innovative technologies under climate-smart agriculture, we not only adapt to climate-induced weather extremes and achieve mitigation of GHGs from agriculture, but also realize sequestration of carbon. There is a huge carbon sink potential in agriculture sector (Corsi et al, 2012).Globally, soil carbon sequestration alone has about 89% of the total mitigation potentialin agriculture sector In other words, agriculture which is a source of GHG emissions, can in turn be converted into sink – through soil carbon sequestration, manure management, Methane trapping from animal dung etc. Biogas plants not only trap Methane from animal dung into fuel, but generated slurry can also be used as manure.

Involving women and local communities in micro-irrigation and watershed management – and linking such community participation to governmental employment guarantee schemes like MGNREGS (Mahatma Gandhi National Rural Employment Guarantee Scheme) – would go a long way in augmenting the water resources for agriculture. Similarly, in the light of unscheduled load shedding of electricity and extreme voltage fluctuations that burn pump sets, installation of solar powered water pump sets would ensure sustained and renewable power supply for irrigation purposes. This will also have other co-benefits in terms of avoided emissions from conventional electricity, economic benefits of renewable energy etc.

Life Cycle Approach to Agriculture: Climate-smart agriculture requires integrated farming system with lifecycle approach at landscape and ecosystem scale, besides many field-based and farm-based sustainable agricultural land management practices such as conservation tillage, agroforestry, residue management, manure management, solar-powered water pump sets etc. Climate-smart agriculture in true sense requires actions beyond the farm scale. One element of FAO's definition is 'adopting an ecosystem approach, working at landscape scale and ensuring inter-sectoral coordination and cooperation'. In the World Bank's version, climate-smart agriculture includes integrated planning of land, agriculture, fisheries, and water at multiple scales, from local level to watershed and regional level.

While decoupling agricultural productivity from the associated environmental degradation and GHG emissions on one hand, the project simultaneously attempts to couple agricultural productivity with:

- Watershed treatment & management,
- Augmentation of renewable energy sources through solar powered pump sets,
- Generation of biogas from animal dung for fuel purpose,
- Community participation & involvement of women,
- Forward and backward linkages for agricultural produce,

- Improving the household incomes of smallholder farmers etc.

Conclusion

Thus Climate-smart agriculture takes a systems-based lifecycle approach to agriculture from community perspective, starting from field/farm based sustainable practices to macro scale institutional and structural interventions at landscape and ecosystem levels.

The objectives of enhancing crop productivity and mitigation of GHG emissions can be achieved only when those objectives are linked to food & livelihood securities and ensured and sustained incomes to local communities. The local community has to be motivated to accept the climate-smart agricultural interventions. Whereas the cost-benefit analysis of complete lifecycle of agriculture favours climate-smart interventions, the benefits cannot be realized in a very short period. This requires some financial assistance mechanism until the local communities can continue CSA on their own, where after the communities can even scale it up without external support.

References

Corsi S, Friedrich T, Kassam A, Pisante M and Sà J. (2012). Soil Organic Carbon Accumulation and greenhouse gas emission reductions from Conservation Agriculture: a Literature Review: UN Food and Agriculture Organization, Rome.

Grainger-Jones, Elwyn. (2012). Climate-Smart Smallholder Agriculture. International Fund for Agricultural Development (IFAD). http://www.ifad.org/pub/op/3.pdf

INCCA (2010). Indian Network for Climate Change Assessment, India: Greenhouse Gas Emissions 2007. Published by the Ministry of Environment and Forests, Government of India, New Delhi. 84p.

Jat R, Sahrawat K, Kassam A. (2014). Conservation Agriculture: Global Prospects and Challenges. Boston, MA, USA: CABI.

NAAS, (2014). Carbon Economy in Indian Agriculture. Policy Paper No. 69, National Academy of Agricultural Sciences, New Delhi: 19p.

NAAS, (2012). Management of Crop Residues in Context of Conservation Agriculture. Policy paper No. 58, National Academy of Agricultural Sciences, New Delhi: 12p.

World Bank, 2014. Building Competitive Green Industries: The Climate and Clean Technology Opportunity for Developing Countries. The World Bank: Washington D.C.

14

Financial Instruments Available in Public and Private Sectors for Climate Smart Agriculture

R. Sudhakar Rao

Former Director of Research, ANGRAU, Hyderabad.

14.1 Introduction

India is home to one third of the world's poorest people, and faces a complicated and integrated set of climate compatible development (CCD) challenges. Although it is a middle-income country, with a current GDP growth rate of 4.8%, 400 million remain under the poverty line. It is ranked 135th on the Human Development Index. Over half of the populations depend on the agriculture sector for their livelihood. India is acutely vulnerable to climate change and its vulnerability is set to increase between 2010 and 2030. Depending on the magnitude and distribution of warming, climate change impact projections for the mid-term (2012-2039) period for India indicated a 4.5 to 9% yield reduction which may roughly amounts to 1.5% of GDP every year. Realizing the impact of climate change, the Government of India has prioritized the research on climate change and a mega project "National Initiative on Climate Resilient Agriculture (NICRA)" has been initiated in 2010-2011.

Agricultural productivity is increasingly affected by changing temperatures, rainfall patterns and natural disasters. It is also the fifth largest GHG emitter in the world and 70% of electricity production is from coal despite a large renewable energy potential. There is therefore a risk of locking in the country to a future high-carbon economy.

State Governments are a crucial actor in responding to climate change challenges. They have an important role in delivering national CCD policies as well as authority to prioritise their own budgetary resources. However, a lack of capacity as well as access to expert information and advice on CCD is limiting their effectiveness.

14.2 Summary of Grant Proposals (Rs. Crores, 2010-2015) by Government of India

Rewarding environmental performance	4000
Building resilience to climate change	2500
Water, sanitation, etc.	35000
Clean energy of cooking	8800
Facilitate solar lantern use	500
Incentivizing Environmental actions	5000
Total	55800

The impact of agriculture on the environment is unquestionable. It is responsible for up to 25% of all greenhouse gas emissions. Policies must be implemented worldwide to help mitigate climate change and raise farmer incomes. Resource poor small farmers live a hand-to-mouth existence. They typically lack the resources to invest in potentially life-changing interventions – even simple one like improved seeds, fertilizers, pesticides, herbicides or improved livestock – or are reluctant to do so because of the risks to their lives and livelihoods if their crops fail or their livestock die.

The difficulties facing farmers are being compounded by climate change. Extremes of weather are increasingly common, making farming a more and more risky business in the immediate term.

The international research programme "Climate change, agriculture and food security" (CCAFS) is an important institution linking climate change and agriculture.

It is not as if climate change has not been discussed at the national policy levels in these countries, most of the South Asian Countries have developed national climate change action plans and have also reported progress to the UNFCCC Secretariat. India, for instance, has the national action plan, initiated by the Prime Minister's Council on Climate Change.

Sri Lanka has a national policy and a national adaptation strategy. The problem is of convergence of the national policies and actions for the farmers.

A US$20 investment in rice research will lift one person out of chronic poverty. R&D in the long run will uplift farmers from poverty and encourage the current and future generations to have greater interest in farming. The international development community needs to understand and appreciate the long-term nature of agricultural research, particularly in response to the challenges of ever-growing populations, diminishing resources, and the impacts of climate change.

Over two-thirds of agricultural land in India is rain fed, as a result of climate change, droughts are increasingly frequent. The major river valleys in the north of the country – the Ganges, Brahmaputra and Indus river systems - have always been prone to flooding, but the area of land affected by floods has more than doubled in recent decades, from 19 million ha in 1950 to 40 million ha in 2002. Between 1801 and 2002, India suffered from 32 serious droughts that reduced agriculture production. In early 2012, parts of western India were suffering from the worst drought in more than 40 years. Indian agriculture has been described as a gamble on the monsoon.

14.3 Impacts of Projected Climate Change on Yield of Different Crops in India

Crop	Region	Yield impact
Rice	All India	Increase (region wise and scenarios based)
	North-West, Central	Increase (decrease in North West)
	South	Decrease/increase (scenarios based)
	North-West	Decrease/increase (scenarios based)
	Punjab	Decrease/increase (scenarios based)
	All India	Decrease
	Northern India	Decrease

Contd...

Crop	Region	Yield impact
Wheat	North	No impact/decrease (scenarios based)
	North-West	+ 16 to + 37%
	North-West, Central	No effect in NW, 10-15% decrease in Central
	North-West	Decrease/increase (scenarios based)
	Punjab	Decrease/increase (scenarios based)
	All India	Decrease
	All India	Decrease/increase (region wise and scenarios based)
	All India	Decrease/increase (region wise and scenarios based)
Soybean	All India	Decrease/increase (region wise and scenarios based)
	Central	Decrease/increase (scenarios based)
Maize	North	Decrease/increase (region wise and scenarios based)
	Punjab	Decrease
Chickpea	All India	Decrease
Pigeon pea	All India	Decrease
Groundnut	Punjab	Decrease
Sorghum	All India	Decrease
	All India	Decrease/Increase (region wise and scenarios based)
Brassica (Mustard)	North	Increase

14.4 Climate Forecasting

For the farmer, climate change is an additional problem that sits on top of all the worries and risks that he has at present. He is not worried about whether the temperature will rise over the next two decades. What he is worried is whether he would get a profitable return from the resources that he has invested on the crop in the field. Since the farmers cannot visualize the long term impacts of climate change, the onus on the experts is threefold. They have to model what direction climate change would take, assess what adverse impact that would have for the farmer, and then develop interventions to deal with them (scientific, policy-level and institutional).

Even as climate forecasting models are getting better, converting predictions for larger areas for smaller locations is still a work in progress.

14.5 Crop Diversification

Mitigating greenhouse gas emissions from agriculture is more difficult, since it has to be done without adversely affecting production. Changing practices such as reduced tilling of the soil and alternate wetting and drying of paddy fields can help reduce emissions. Research is ongoing to develop livestock breeds that produce less methane and also developing fodder additives that can reduce methane generation.

Though it is certain that climate is changing and extreme weather events are becoming more frequent, it is still uncertain how much and in what fashion it will change. The challenge for research, policy and implementation is to remain one step ahead of the process so that farmers' livelihoods and national food security are not compromised.

14.6 Impact on Water Resources during the Next Century Over India

Region/location	Impact
Indian sub-continent	Increase in monsoonal and annual run-off in the central plains
	No substantial change in winter run-off
	Increase in evaporation and soil wetness during monsoon and on an annual basis
Indian Coastline	One metre sea-level rise on the Indian coastline is likely to affect a total area of 5763 sq kms and put 7.1 million people at risk
All-India	Increase in potential evaporation across India
Central India	Basin located in a comparatively drier region is more sensitive to climate changes.
Kosi Basin	Decrease in discharge on the Kosi River
	Decrease in run-off by 2-8%

Contd...

Region/location	Impact
Southern and Central India	Soil moisture increase marginally by 15-20% during monsoon months.
Chenab River	Increase in discharge in the Chenab River
River basins of India	General reduction in the quantity of the available un-off, increase in Mahanadi and Brahimini basis
Damodar Basin	Decreased river flow
Rajasthan	Increase in evapo-transpiration

14.7 Co-ordinating Finance for Climate Smart Agriculture

Investing in climate-smart agriculture at a landscape scale will have a large price tag. For example, in order to achieve food security for growing population, an estimated net US$83 billion a year will be required in developing countries and US$11 billion in Sub-Saharan Africa alone. In general, the World Bank estimates that mitigate measures in developing countries could cost between US$140 - $ 175 billion per year for the next twenty years.

14.8 Barriers to Coordinated Financing for Climate-smart Agriculture

The analysis of climate and agricultural development funds did yield powerful insights about the barriers to streamline and scale-up funding for climate-smart agriculture, although estimating the total scale with a high level of precision proved to be quite challenging. Insights from the data include: (1) International public funding sources are uncertain, (2) Climate finance is fragmented and (3) Private agriculture investments are the main drivers of land use decision, although climate finance would be substantial if international funding commitments were honoured; and (4) Public funds supporting climate action and those supporting agriculture remain largely separate.

14.9 Huge Financing Gap

Considerable investment in filling data and knowledge gaps, research and development of appropriate technologies, and incentives to ensure adoption of climate-smart agricultural practices is needed. FAO's report says:

Funding also should be targeted towards rebuilding neglected national agricultural extension services, which will have a key role to play in supporting farmers as they transition to climate-smart agriculture.

But FAO warns that currently insufficient resources are available for financing efforts to help agriculture and farmers prepare for climate change especially in the developing world.

"Climate change will increase the overall investment requirements needed to achieve food security, but financing resources currently available are substantially insufficient" and "climate financing – both existing and that under discussion – does not take explicit account of the specific requirements of developing country agriculture, "its report says., it is unlikely that public or private resources alone will suffice, innovative ways of blending these resources will challenge financial mechanisms.

The report cites World Bank estimates for the annual costs of climate change adaptation in developing world agriculture of $ 2.5 – 2.6 billion per year between 2010 and 2050, as well as the UNFCC estimate the additional investment and financial flows needed in developing countries for mitigation in the agricultural sector of $14 billion annually by 2030.

14.10 Better Policies, Stronger Institutions

FAO's report also argues that greater coherence among agriculture, food security and climate change policy-making is urgently needed.

"Policies in all three of these areas impact small holder production systems and a lack of coherence can prevent them from capturing synergies", it says, stressing the need to establish mechanisms that allow dialogues between policy makers working in these areas.

Improving mechanisms for getting data, science and know-how to farmers so they can adapt is another area in need of attention.

Agricultural Extension Systems

Have in the past been a key conduit for disseminating information and knowledge to farmers, but in many developing countries these systems having long been in decline, the report warns. The Farmers Field School system pioneered by FAO offers an additional channel for promoting knowledge transfer and adoption of climate-smart farming techniques.

Additionally, the report notes that effective systems of use and access and property rights are essential to improving management of natural resources.

And new types of accessible and affordable insurance that can help farmers weather the impact of climate change need to be explored.

14.11 Climate Change has Critical Ramification for the Country's Food Security

Knowledge about the impact of climate change on current water and crop production is limited. At the same time mitigating and brining a halt to climate change is not within the capability of one country alone. Thus adaption strategies seem to be the most immediate needs to save livelihoods and ensure food security.

India has to maintain the sustainability of its ecosystems to meet the food and non-food needs of a growing population. The main thrust of the programmes to combat the impact of climate change on food security should be activities relating to rainwater harvesting and soil conservation.

Public investment in agriculture has fallen dramatically since 1980s. This coincides with declining share of agriculture in the total gross capital formation (TGCF). Instead of promoting low-cost labour intensive options that have a higher capital-output ratio, present policies have resulted in excessive use of capital on the farms such as too many tube wells in water-scarce region.

Another big change in the last three decades is the dominant use of groundwater as opposed to surface and sub-soil (though shallow wells). Groundwater has become the main source of irrigation. Surface irrigation systems already created are lying wasted because canals or other systems are hardly maintained. Because of inefficiency of large water irrigation systems, people have been forced to exploit groundwater. Thus bulk of Indian agriculture not only remains rain fed but also depends on groundwater, not surface water. This is worrisome in the current context of increasingly variable.

14.12   Climate Change, a Crisis Catalyst

The rate of CO_2 release into the atmosphere has increased by 30 times in the last three-four decades. It is estimated that a 0.5 degree Celsius rise in winter temperature would reduce wheat yield by 0.45 tonne per hectare. A recent World report studied two drought prone regions in Andhra Pradesh and Maharashtra and one flood prone region in Orissa on climate change impacts. It found that climate change could have the following serious impacts:

- In Andhra Pradesh, dryland farmers may see their incomes plunge by 20 percent.

- In Maharashtra, sugarcane yields may fall dramatically by 25-30 percent.

- In Orissa, flooding will rise dramatically leading to a drop in rice yields by as much as 12 percent in some districts.

Other effects of climate change are more pronounced. For instance, rise in sea levels, say about a meter by the next century, may displace millions of people. Sea level rise would lead o ingress of saline water and salination of ground water and surface water in coastal areas. Salt water intrusion in low-lying agricultural plains could lead to food insecurity, further spread of water-related diseases and reduced freshwater supplies.

Driving Forces of Assessment at the Vulnerability of Indian Agricultural Production to Climate Change

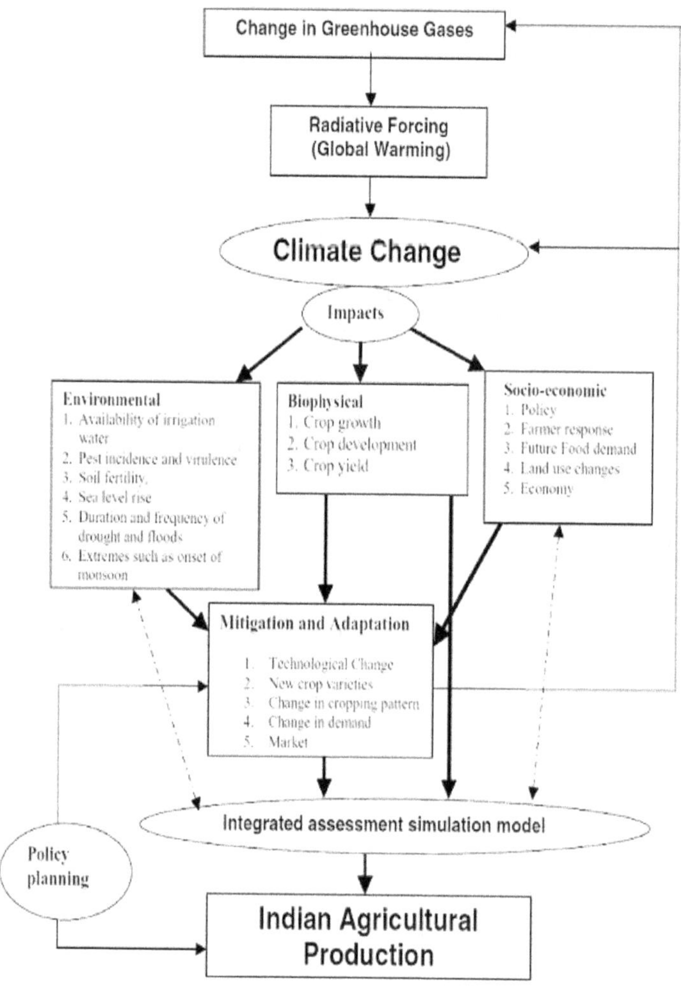

14.13   Drought-Tolerant Maize Boosts Food Security for Millions of African Farmers

Maize is a staple food for more than 300 million people in Africa but, by the 2030s, drought and rising temperatures could render 40% of the continent's current maize-growing area unsuitable for maize varieties available today. Maize production in southern Africa, for example, may fall by 30% or more. New, drought and heat-tolerant varieties will have to be developed quickly and be growing in farmers' fields in the next few years if we are to avoid widespread famine in Africa.

Since 2006 more than 100 new, drought-tolerant maize varieties and hybrids have been developed and released across 13 countries by the Drought Tolerant Maize for Africa Initiative (DTMA), funded by the Bill & Melindia Gates Foundation, the Howard G.Buffet Foundation, USAID and the UK Department for International Development (DFID).

Key to the success of this initiative, which is coordinated by the International Maize and Wheat Improvement Centre (CIMMYT) and the International Institute of Tropical Agriculture (IITA) is the way it has brought together a wide range of partners, including publicly-funded research organizations, public and private seed producers, varietal certification agencies and farmer groups.

14.14   Herbicide-tolerant Crops Contribute to Climate Change Resilience and Mitigation

Herbicide – tolerant (HT) and pest-resistant crops boost the climate resilience of farming systems and their capacity to mitigate climate change. HT crops, for example, reduce the need of ploughing and other types of mechanized weed control, reducing fuel consumption by upto 44% in maize and 60% in soybean. Both HT and pest-resistant crops reduce the amount of chemicals farmers need to apply, and the chemicals they use are less toxic than previous generations of herbicides and pesticides. Reduced tillage help to preserve soil structure, reducing erosion and increasing infiltration and retention of water, and leads to a built up of organic matter in the soil. Such benefits protect the environment and increase the resilience of farming systems, while also reducing the contribution of agriculture to climate change.

14.15   High Hopes for GCF Green Climate Fund (GCF)

One attempt to make climate funds more transparent and easier to access is through the Green Climate Fund (GCF), jointly managed by the UNFCCC and the Global Environment Facility (GEF). In a transitional phase until 2013, the GCF is anticipated to provide recipient countries direct access to climate change finance, through accredited national implementing entities in addition to multilateral agencies. Pledges for the GCF and the associated 'fast start' financing have been made for US$30 billion by 2012 and US$100 billion a year by 2020 to be balanced between adaptation and mitigation. The criteria for the allocation of these funds are still unclear, as is the distribution between public and private sources, and the level of funding that will ultimately materialize. From the promised US$30 billion a year of fast-start funds, only 8% has been disbursed, in many cases drawing from development aid funds.

Private agriculture investments are the main driver of land use decisions; although climate finance would be substantial if international public funding commitments were honoured:

Private agriculture investment throughout the world is on the rise. This is particularly true in Africa with the recent rise in FDI and land acquisitions, where there are billions of dollars per year flowing towards agriculture production and management in Africa. If the full agricultural value chain is included, investment could approach a hundred billion dollars, or more per year. Climate funds have the potential to grow and become more significant if the GCF is fully funded in the future – US$100 billion per year globally – and agriculture begins to play a more significant role in carbon markets and NAMAs. However, the extent to which this happens will largely be a question of policy decision made over the next few years.

Private investment will continue to rise in the future, as indicated by India's growing low carbon environment market, and the increasing number of environmental and green funds

Private finance investment varies significantly across all sectors in India

India has seen approximately USD 19 billion of investment in the **Renewable Energy (RE)** sector since 1999. A combination of financial sources and instruments were utilised, however the majority has been a mixture of local and international private investors financing initiatives in the form of debt

Energy efficiency (EE) investments (within the building technology, the industry applications and part of the 'Mixed' category) have accounted for approximately USD 3 billion of private finance in India since 1994. The majority, 2 USD billion, of finance has been provided by multilateral development banks (World Bank and ADB) in the mixed EE category.

The **Transport sector** has had an investment of approximately USD 4.6 billion; supporting initiatives in modal shift from road to rail/long distance rail infrastructure; modal shift to public transport and road infrastructure planning and improvements

Private equity being invested in the **Cleantech sector** since 2005 totalled USD 3.6 billion from 116 deals (VCCEdge, n.d.); unfortunately a breakdown of this information was not available due to confidentiality and commercial reasons.

Since 2006 approximately USD 174 million of private finance has been invested in the **Waste Management sector**. This encompasses investment in waste services (municipal and industrial), waste to energy, e-waste management, solid waste processing and recycling.

The **forestry sector** received USD 1.8 billion worth of investment from JICA since 1991 directed towards India's national Afforestation Program. This was provided in the form of ODA loan assistance, and therefore is not strictly private investment.

India's Low Carbon Environmental Goods and Services (LCEGS) market in 2011-12 was USD 323 billion.

It was ranked 4th in the world, behind USA, China and Japan respectively. Furthermore, in BNEF's Climates cope 2014 report, which evaluates climate-related investment on a country-by-country basis, India was ranked 4th out of 55 countries reviewed and it had its best performance on Low Carbon Business and Clean Energy Value Chain parameters.

The estimated total market value of all the financial assets under management (AUM) by the environmental and green funds in India is currently approximately USD 222 billion.

There are an increasing number of environmental and green funds in India, primarily investing in renewable energy companies, technology providers and infrastructure sectors. The majority of funds reviewed in this study were managed by venture capital firms.

The Renewable Energy Market in India is anticipated to see investments worth USD 200 billion by 2022.

India has been able to achieve only 13% of its renewable energy potential, which as of March 2014 was 216, 918 MW. The planned USD 100 billion investment to develop India's solar market would increase the capacity to approximately 175 GW by 2022.

The environmental technologies market is currently estimated to be worth approximately USD 9 billion per year, and is expected to grow by 15% annually.

In addition to this, it is estimated that the industrial energy efficiency market potential in 2018 will be around USD 27 million.

The water and waste management sectors have investment plans of USD 50 billion over the next five years.

The Indian municipal solid waste (MSW) management market is expected to grow at a Compound Annual Growth Rate (CAGR) of 7% by 2025 while e-waste management market is expected to grow at a CAGR of 10.03% during the same period.

There are numerous technical and knowledge barriers leading to a lack of a pipeline of bankable projects

There is limited understanding of climate risks in private sector investments. Private sector engagement in climate change has been limited to greenhouse gas (GHG) accounting, and there has been no use of robust climate risk screening tools by the private sector.

- Technical knowledge of low carbon sectors to increase investments from FIs perspective have mostly been driven by senior management; projectOfficers appraising the project do not have the requisite knowledge of low carbon sectors.

- Significant scale-up of capacity and removal of barriers are still required to create a strong pipeline of investible projects. Bankable projects are increasingly being developed but mainly in the RE sector. Developing a strong pipeline of bankable projects in the EE sector is still a big challenge.

- Several barriers such as high transaction costs, small project size, lack of EE understanding amongst financial institutions, etc. have hindered the process.

- Financing energy efficiency projects is more risky due to the small size of the projects and dispersed nature of the technologies. Indian FIs (such as

- TCCL and ILFS) do not currently have the capacity to cater for the project sizes the Energy Service Company (ESCO) market generates.

- Project developers and financial intermediaries find eligibility criteria and reporting requirements of donors to be extremely onerous. International donors have onerous access and reporting requirements and lengthy project application timescales; which disincentives Indian project developers.

Public funds supporting climate action and those supporting agriculture remain largely separate:

In addition to growing private sector investment in agriculture, public sector funds are growing as well, particularly in Africa where institutions and investments plants are developing quickly. These public investments can support the public goods provided by agriculture, which include not only increase in production, but other potential benefits including poverty reduction, the provision of environmental services and even climate change adaptation and mitigation. With this broad view of the role of agriculture, the line between public investment in agriculture, climate, and sustainable and land management would begin to blur.

National climate policies are showing positive signs for scaling up private climate finance

The National Mission on Enhanced Energy Efficiency (NMEEE) and the National Solar Mission (NSM) are showing strong potential to engage and leverage private finance by addressing key barriers

	NMEEE	NSM
Objective	The NMEEE promotes innovative policy and regulatory regimes, financing mechanisms, and business models to promote energy efficiency	The NSM aims to achieve grid parity for solar electricity through research & development, domestic production, large scale deployment, and long term and predictable policy that encourages private sector participation in the solar business
Institutional / Political	Setting up of committees for implementation of PAT, PRGFEE and VCFEE A robust framework for implementation of the PAT, using highly consultative process Initiation of new programmes e.g. MTEE's Super-Efficient Equipment Programme (SEEP) Creation of EESL (super ESCO) as a corporate entity to provide market leadership under EEFP Assigned state nodal agencies for implementation of the Mission at the state level	FITs, RECs, RPOs, Power bundling, Accelerated depreciation benefit, Asset liability mismatch and non-availability of long term debt is acknowledged by lenders Stricter enforcement of RPOs to steer the market further.
Financial / Economic	Designed and soon to be implemented new financial mechanisms, in the form of partial risk guarantee fund and venture capital fund, building the confidence of FIs to invest in EE	AD benefit, CDM, MDB financing, bank guarantees, Renewable Energy Infrastructure Development Fund (REID), off-grid fund, viability gap funding (VGF), credit guarantees, charge exemptions. Generation Based Incentives, Manufacturing incentives, Off grid incentives Project financing by IREDA and PFC
Technical / Knowledge	Preparing bankable project DPRs, to build capacities of ESCOs to seek financing from financial institutions Standardizing ESCO performance contracts, to assist ease of understanding of such contracts by project implementers and also financiers Developing performance measurement and verification tools, tailored for Indian EE projects	Creation and up-gradation of Solar Energy Centre for R&D Setting up of Solar Energy Corporation of India (SECI) for facilitating financing, training and capacity building, enhancing technical know-how on solar Supporting a number of Centres for R&D in different institutes
Private Sector Engagement	Sub programmes are primarily designed to encourage private sector investment and finance in the various programmes on EE in India by developing a project pipeline and new financing instruments	The NSM been very successful in attracting private sector investment by creating new financing, implementation and regulatory structures

Budget allocated for AICRPAM-NICRA during 2014-15

S.no.	Name of the Centre	Contingency (Rs)	TA (Rs.)	Total (Rs.)
1.	Akola	800000	30000	830000
2.	Anand	950000	40000	990000
3.	Anantapur	930000	30000	960000
4.	Bangalore	1000000	30000	1030000
5.	Bhubhaneswar	1405247	40000	1445247
6.	Bijapur	1030000	30000	1060000
7.	Chatha	1006000	30000	1036000
8.	Dapoli	750000	30000	780000
9.	Faizabad	680000	24000	704000
10.	Hisar	700000	25000	725000
11.	Jabalpur	830000	30000	860000
12.	Jorhat	1210000	40000	1250000
13.	Kanpur	1150000	25000	1175000
14.	Kovilpatti	1230000	55000	1285000
15.	Ludhiana	1109000	21000	1130000
16.	Mohanpur	720000	30000	750000
17.	Palampur	1250000	30000	1280000
18.	Parbhani	830000	10000	840000
19.	Raipur	950000	35000	985000
20.	Ranchi	730173	30000	760173
21.	Ranichauri	800000	25000	825000
22.	Samastipur	839580	25000	864580
23.	Solapur	1200000	40000	1240000
24.	Thrissur	1000000	50000	1050000
25.	Udaipur	900000	45000	945000
	Total	24000000	800000	24800000

National Mission for Sustainable Agriculture

14.16   Mission Objective

To transform agriculture into an ecologically sustainable climate resilient production system while at the same time, exploiting its fullest potential and thereby ensuring food security, equitable access to food resources, enhancing livelihood opportunities and contributing to economic stability at the national level.

Mission Targets and Timeline

To achieve its objective, the mission will work on the following major programme components or activities:

- Rainfed Area Development: Adopt an area based approach for development and conservation of natural resources along with farming systems

- On-Farm Water Management: Enhance water use efficiency by promoting efficient on-farm water management technologies and equipment

- Soil Health Management: Promote location as well as crop specific sustainable soil health

- management

- Climate Change and Sustainable Agriculture, Monitoring, Modeling and Networking:

Creation and bidirectional (farmers to research institutions and vice versa) dissemination of climate change related information and knowledge

Budgetary Requirements and Allocations

The mission requires budgetary support of INR 1,08,000 crore (approx. USD 17.4 billion) up to the end of 12th five year plan period (2011 2017). Proposals for INR 13,034 crore (approx. USD 2.1 billion) have been approved.

Implementation Status

Key achievements to date:

- Developed 11,000 hectares of degraded land

- 1 million hectares brought under micro-irrigation to promote water efficiency

- Created 5.4 million metric tonne agricultural storage capacity

References

Beddington, J.R., Asaduzzaman, M., Clark, M.E., Fernández Bremauntz, A., Guillou, M.D., Jahn, M.M., Lin,E., Mamo, T., Negra, C., Nobre, C.A., Scholes, R.J., Sharma, R., Van Bo, N. & Wakhungu, J. (2012c). The Role for Scientists in Tackling Food Insecurity and Climate Change. Agriculture and Food Security, 1: 10.

Buchner, B., Falconer, A., Hervé-Mignucci, M., Trabacchi, C. & Brinkman, M. 2011. The Landscape of Climate Finance: a Climate Policy Initiative (CPI) report. (available at http://climatepolicyinitiative.org/wp-content/uploads/2011/10/The-Landscape-of-Climate-Finance-120120.pdf).

Conforti, P., eds. (2011). Looking Ahead in World Food and Agriculture: Perspectives to 2050. Rome, FAO.

European Commission. (2013). Other Sources of Climate Financing. (available at http://ec.europa.eu/clima/policies/finance/international/other/index_en.htm).

FAO. (2010). "Climate-smart" Agriculture: Policies, Practices and Financing for Food Security, Adaptation and Mitigation. Rome.

Füssel, H.-M. & Klein, R.T.J. (2006). Climate Change Vulnerability Assessments: an Evolution of Conceptual Thinking. Climatic Change, 75(3): 301–329p.

Gitz, V. & Meybeck, A. (2012). Risks, Vulnerabilities and Resilience in a Context of Climate Change. In FAO & OECD Building Resilience for Adaptation to Climate Change in the Agriculture Sector. Rome.

HLPE. (2012b). Social Protection for Food Security. A report by the High Level Panel of Experts on Food Security and Nutrition of the Committee on World Food Security, Rome.

Huang, H., von Lampe, M. & van Tongeren, F. (2011). Climate change and trade in agriculture. Food Policy, 36: S9-S13.

International Finance Cooperation (IFC). (2011). Climate Finance: Engaging the Private Sector.

Meridian Institute. (2011). Agriculture and Climate Change: a Scoping Report. (available at http://www.climateagriculture.org/Scoping_Report.aspx) Meybeck, A., Azzu, N., Doyle, M. & Gitz, V. 2012. Agriculture in National Adaptation Programmes of Action.

Padgham, J. (2009). Agricultural Development under a Changing Climate: Opportunities and Challenges for Adaptation. Washington D.C., The World Bank.

Shames, S., Friedman, R. & Havemann, T. (2012). Coordinating Finance for Climate-Smart Agriculture. Eco agriculture Discussion Paper No. 9, Eco agriculture Partners. (available at http://ecoagriculture.org/documents/ files/doc_431.pdf).

United Nations Environment Programme (UNEP). (2010). Bilateral Finance Institutions and Climate Change: a Mapping of 2009 Climate Financial Flows to Developing Countries.

World Bank. (2012). Inclusive Green Growth: the Pathway to Sustainable Development. Washington D.C, The World Bank.

15

Risk Transfer Mechanism in Climate Smart Agriculture

G. Kiran Kumar
Centre for Good Governance, Hyderabad.

15.1 Introduction

Vulnerability to climate change and other risks can increase over time, and potential gains on the path towards more resilient, productive and sustainable livelihoods can be at risk if households face repeated shocks that steadily erode their assets and with that their resilience to future shocks. Resource transfers can play a role in preventing this from happening by providing cash, vouchers, food or other resources during periods of crisis and during the cyclical 'hungry season' (WFP, 2011a)[1]. Resource transfers can prevent seasonal malnutrition among vulnerable populations; reduce debt and increase savings and productive investment, even in areas that are subject to recurrent crises (FAO, 2012). They provide poor households with an alternative from having to revert to so-called "negative coping strategies" that leave them more vulnerable. These include selling their productive assets—such as farm tools or livestock— to buy food, taking children out of school (which has irreversible impairments to their long-term socio-economic potential and resilience) or over-exploiting lands and forests (WFP, 2011a). Resource transfers, conditional and unconditional, have in fact been an important and well-performing part of the response to major natural disasters in the past (Heltberg et al., 2010)[2].

---

[1]FAO. 2011a. Evaluation of supporting conservation agriculture for sustainable agriculture and rural development (CA for SARD). Phase II report.

[2]Heltberg, R., Siegel, P. B. & Jorgensen, S. L. 2010. Social policies for adaptation to climate change, in social dimensions of climate change: equity and vulnerability in a warming world, R. Mearns & A. Norton, eds. Washington D.C., The World Bank.

Safety nets play a critical role in protecting human, social, physical, financial and natural capital, which are the foundations of resilient and productive livelihoods (HLPE, 2012b; PHI et al., 2011; Davis, 2011; WFP, 2011a). This is both relevant for protecting the poorest and the most food insecure people and communities from falling into destitution, and for protecting those who are better off and already moving towards more resilience, productivity and sustainability from falling (back) into poverty. In this regard, safety nets could also protect potential gains and achievements in CSA.

Agriculture is inherently risky, and may be even more so in the future with more extreme climate events. For poor farmers, adopting new technologies and production strategies may be beyond their tolerance for risk, given that failure may be catastrophic (FAO et al., 2012)[3]. It also often requires a certain investment, which – even if minimal – may be beyond their capacity. This applies also to many climate-smart production strategies. As the HLPE Report on Climate Change and Food Security stresses, financial capital for investment in adaptation as well as human and social capital to implement adaptation are major obstacles for poor farmers (HLPE, 2012b)[4]. Many innovations that could make them more resilient, for example weather-index based insurance, are therefore difficult to access by those in greatest need. If market failures in credit, insurance and other areas are not addressed, households will be limited in their ability to adapt to climate change – or manage any other risk (Davis, 2011)[5]. Safety nets offer significant potential to meet these challenges, which is often underestimated, as they continue to be predominantly viewed as a last resort "when adaptation fails" (HLPE, 2012b) – certainly also because of misleading connotations with the term "safety net". However, there is increasing

---

[3]FAO. 2012. FAO policy on cash-based transfers. Rome.

[4]HLPE. 2012b. Climate change and food security. A report by the CFS HLPE. Rome.

[5]Davis, B. 2011. Cash transfer programs in SSA: recent developments and implications for climate change adaptation. Presentation for session on Building resilience to climate change in smallholder agricultural systems, FAO ICABR, Frascati.

recognition of the contribution of safety nets to building resilience, reducing risks and enhancing adaptive capacity among vulnerable communities. Various empirical studies and impact evaluations have indeed shown that safety nets can spur economic growth among poor communities, which relates to the productivity element of CSA. The role of social protection and safety nets in making growth inclusive has placed them firmly in the discussions on sustainable development and on an "inclusive green economy" By providing a basic level of consumption below which people know they cannot fall, and by boosting a household's asset base, resource transfers can allow poor farmers to invest further in education, health, skills development and productive assets. All of these are not only prerequisites for productive and sustainable livelihoods, but also critical elements of adaptive capacity and resilience (UN TT SDCC, 2011)[6].

In addition, public works programmes that guarantee employment when needed, effectively provide insurance. Agriculture-related public works activities, such as hillside terracing or soil and water conservation, can improve farm yields and generate sustainable benefits for household food security. They can also create community assets and infrastructure critical for adaptation. Studies of the long-term impacts of natural resources management activities undertaken through food-for-work projects (in particular those that included terracing, agroforestry and water capture and spreading) have found significant impacts in terms of increased crop yields as well as increases in vegetation diversity and cover.

15.2 National Agricultural Insurance Scheme (NAIS)

The Government of India introduced the scheme from Rabi 1999-2000 season to protect the farmers against losses suffered by them due to crop failure on account of natural calamities. The scheme is currently implemented by Agriculture Insurance Company of

---

[6]United Nations Task Team on Social Dimensions of Climate Change (UN TT SDCC). 2011. The social dimensions of climate change. Discussion draft. (available at http://www.who.int/globalchange/mediacentre/ events/2011/social-dimensions-of-climate-change.pdf)

India (AICIL). The scheme is available to all the farmers, loanee and non-loanee, irrespective of size of their holding. The scheme covers all food crops (cereals, millets and pulses) and oil seeds and Annual commercial/ horticultural crops. At present, 10% subsidy on premium is available to small & marginal farmers. NAIS is presently being implemented in 24 States and 2 Union Territories except in States of Punjab & Arunachal Pradesh. Nagaland has given consent to implement the scheme and Rajasthan has decided to implement WBCIS in place of NAIS. Since the inception of the scheme and until up to 31.03.11 about 176 million farmers have been insured, covering an area of 269 million hectares for a sum insured value of Rs. 2,21,213 crore, against a premium of Rs. 6589 crore. Claims to the tune of about Rs. 22190 crore have been reported so far benefiting nearly 47.6 million farmers representing a claim ratio of 1:3.37.

Claims are automatically calculated based on shortfall in the current season yield obtained from crop cutting experiments conducted by State Governments under General Crops Estimation Survey (GCES) as compared to threshold yield and settled through the rural banking network. The Company is making efforts to bring the remaining States/UTs into the fold of NAIS.

Modified National Agricultural Insurance Scheme (MNAIS) will be implemented in 50 selected districts of India on pilot basis in place of National Agricultural Insurance Scheme (NAIS). Agriculture Insurance Company of India Ltd. (AIC) and approved private sector insurance companies mainly engaged in agriculture insurance business have been empanelled by GOI for implementation of pilot MNAIS. Selection of insurance companies among empanelled insurance company will be done by concerned State Government for implementation of MNAIS, in their States. In case the State wishes to use services of more than one insurance company, State Government will ensure that only one insurance company operates in each district.

Conclusion

Safety nets can also serve as a platform for enhancing access to innovative risk management tools, such as weather-index based insurance. For example, "insurance for work" can be used not only to expand access to insurance, but can be added to existing labour-based safety nets to protect beneficiaries and reduce costs for governments and donors from the disruptions caused by climate disasters.

Another area where safety nets can make a significant contribution is in fostering human and social capital, which set the stage for and maximize the impact of adaptation interventions (UN TT SDCC, 2011). Investments in early childhood nutrition can spur economic growth, as these investments have long-term effects on cognitive skills and productivity. Safety nets can also provide a platform for introducing specific CSA related activities – for example capacity development on skills and adaptation practices and technologies to be applied in homestead vegetable production or agro-forestry – and for targeting these activities towards the poorest and most vulnerable, who are often women.

References

http://www.fao.org/3/a-i3325e.pdf

http://financialservices.gov.in/insurance/gssois/nais.asp

http://agricoop.nic.in/mnaiso29910.pdf

16

Planning and Preparation of Projects for Climate Smart Agriculture under NMSA

K. Padmaja
Centre of Good Governance, Hyderabad.

16.1 Introduction

National Mission on Sustainable Agriculture is one among the eight missions envisaged by Government of India (GoI) during XII plan with an objective of achieving sustainable agriculture growth with changing scenarios of climate under the National Action Plan on Climate Change (NAPCC). India like many other countries is facing the threat of climate change for sustainable agriculture growth and development. The need of the hour is to develop a strategy to face the challenges posed by the climate change which in turn affects the national food security and economy. Lack of proper adaptation and mitigation measures would result in adverse impacts on agriculture production which has the consequences of shortages of food grains and rising prices which could endanger the food and livelihood security of our country. Though the NAPCC has given more thrust to the rainfed areas, NMSA mission document has taken care of the coastal and irrigated regions along with rainfed regions of the country.

As per UNDP 2009, climate change is no more an environmental concern. It has emerged as the biggest developmental challenge for the earth. NAPCC drawn the attention of the policy makers that if no adaptation and mitigation measures are taken, we could lose as much as 40% agricultural yields by 2100. Hence it is imperative to have future projections and subsequent fund allocations to encounter the threat.

The NMSA has been established with an aim to have a transformation from current agriculture practices to climate resilient production systems in the domain of crops and livestock through appropriate adaptation and mitigation measures. The mission has given prime importance to rainfed agriculture by acknowledging its vast potential and key role in nation's food security. It intends to meet the objective through developing drought resistant and pest resistant varieties and strengthening institutional support to dry land farmers.

A. Vision of NMSA:

- Transform Agriculture into Climate Resilient Production system

- Grow and Ecologically Sustain agricultural production to its Fullest Potential

- Ensure Food Security and Equitable Access to Food Resources,

- Enhance Livelihood Opportunities, Contribute to Economic Stability at the National Level

B. Objectives of the mission:

- To make agriculture more productive, sustainable, remunerative and climate resilient by promoting location specific Integrated/Composite Farming Systems;

- To conserve natural resources through appropriate soil and moisture conservation measures;

- To adopt comprehensive soil health management practices based on soil fertility maps, soil test based application of macro & micro nutrients, judicious use of fertilizers etc.;

- To optimize utilization of water resources through efficient water management to expand coverage for achieving 'more crop per drop';

- To develop capacity of farmers & stakeholders, in conjunction with other ongoing Missions e.g. National Mission on Agriculture Extension & Technology, National Food Security Mission, National Initiative for Climate Resilient Agriculture (NICRA) etc., in the domain of climate change adaptation and mitigation measures;

- To pilot models in select blocks for improving productivity of rainfed farming by mainstreaming rainfed technologies refined through NICRA and by leveraging resources from other schemes/Missions like Mahatma Gandhi National Rural Employment Guarantee Scheme (MGNREGS), Integrated Watershed Management Programme (IWMP), RKVY etc.; and

- To establish an effective inter and intra Departmental /Ministerial co-ordination for accomplishing key deliverables of National Mission for Sustainable Agriculture under the aegis of NAPCC.

To promote sustainable practices and to face the challenges posed by climate change, this mission identifies ten key dimensions covering both adaptation and mitigation measures which will be implemented through a Programme of Action(POA). This would be carried out through four functional areas viz.

1. *Research and Development, Technologies, products and practices, Infrastructure and Capacity building*. Though the modern technologies played a key role in mitigating climate change, this mission is given the same priority in harnessing the local traditional knowledge for in-situ conservation of genetic resources.

The mission has been designed by converging all the existing ongoing as well as proposed new schemes related to sustainable agriculture on soil & water conservation, water use efficiency, soil health management and rainfed area development. NMSA aims at promoting location specific improved agronomic practices by considering above aspects. By keeping the objectives in view the implementing agencies can prepare projects with customized, location specific interventions.

As this mission aims to converging the existing programmes and schemes under ministry of agriculture and also converging different related ministries, the ministries like ministry of rural development, water resources, environment etc are to be taken into confidence. For better understanding of the schemes and programmes that are

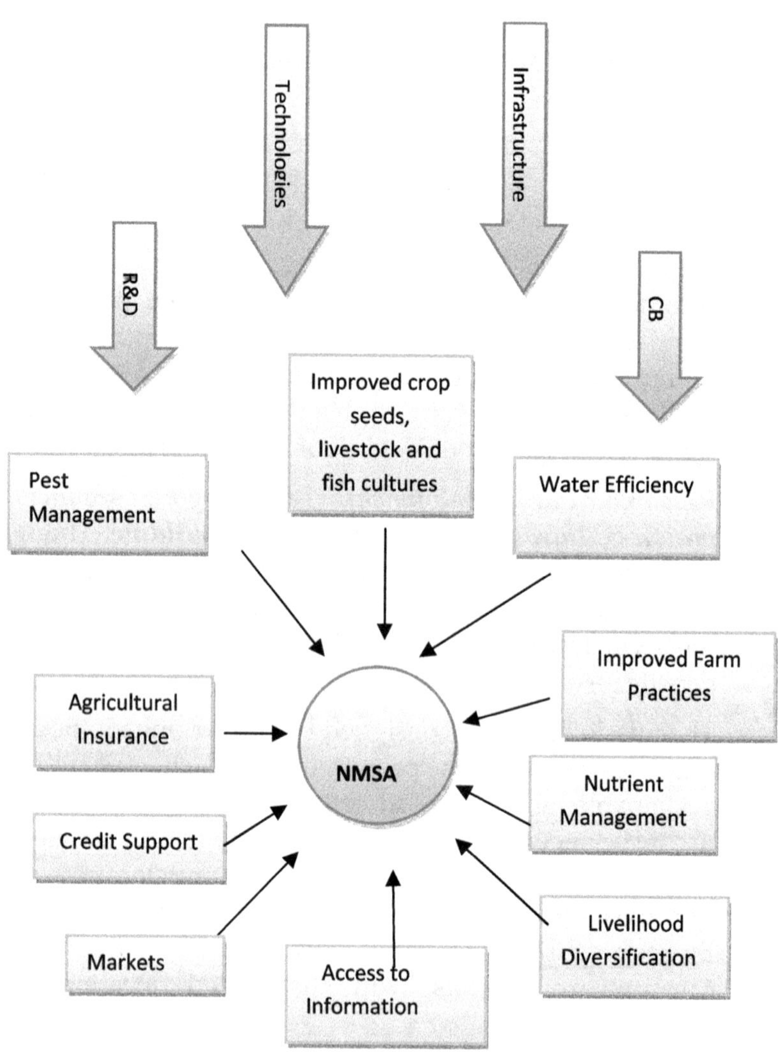

Key Dimension of NMSA

aiming for sustainable agricultural practices like soil and water conservation measures, organic farming, etc which are designed in tandem with climate resilience approach, a few are briefed below. Among these existing interventions there is a scope to upscale most of them.

Table 16.1 Existing interventions being implemented by different ministries of GOI

Functional Area	Intervention	Implementing scheme/mission
Research and Development		
Genetic resources – conservation and development of varieties of crops, livestock, fodder and fisheries	Extending support to research institutions for production of breeder, hybrid and high yielding varieties for crop	MMA, RKVY, NADP and NHM
	Identification and tagging of superior cattle germplasm	NPCBB I and II
	improvement of quality fodder seeds	Fodder Development Scheme
Technology, Products and Practices		
Supply of seed or plant material, seed treatment material,	Support for seeds, planting materials and agricultural inputs including seed treatment	
Resource Conservation	procuring resource conserving implements or machinery at subsidized rate	MMA, RKVY, NFSM
	Distribution of drip and sprinkler irrigation infrastructure for high water consumptive crops and horticulture	NAIP, MMA, MI
Resource (land and water) Remediation technologies and practices	Reclamation of saline and alkali soils	MMA
	Development of degraded and waste lands is done under watershed programmes	

Table 16.1 *Contd...*

Functional Area	Intervention	Implementing scheme/mission
	Precision farming and conservation agriculture	
Organic farming	Promoting organic farming through supply of bio products, green manuring, Vermi composting etc	NPMSF, NADP and MMA
Infrastructure		
Irrigation (Coverage and efficiency – surface and groundwater)	Renovation, desilting of existing infrastructure	NPRRR, NREGS and SGRY
	Promotion of Micro irrigation infrastructure	
Market Infrastructure	Establishment of modern terminal markets in important urban centres	Terminal Market Complex (TMC) scheme
	construction of rural godowns	Grameen Bhandaran Yojana
Insurance	Implementing crop insurance scheme to safeguard the farmers in the event of natural calamities	National Agriculture Insurance Scheme
Credit	Extending support in the form of crop loans, farm mechanization, land development, construction of storage structures etc..	
Capacity Building		
Demonstrations and Trainings	Front line demonstrations for new released crop varieties or new agronomic practices	MMA, NFSM, NADP
	Training through Farmers Field schools to farmers, Season Long Training Programmes for Agrl extension functionaries	MMA
	SAMETI's, MANAGE, EEI are extending training to senior and middle level officers	

C. Operational strategies to develop climate resilient projects under NMSA:

To achieve the climate resilient agriculture, the adaptation and mitigation practices or technologies are to be mainstreamed. They are to be brought into the developmental path ways. This could be achieved through adaptive Research and Development, imbibing global best practices or technologies, and providing improved and adaptive infrastructure not only in terms of physical structures like creating rural godowns, roads etc, but it also covers institutional credit, insurance as a safety net. In addition to those three strategies, customized capacity building to staff of agriculture and allied sectors and farmers is also necessary. Hence the Programme of Action (PoA) of this mission would work under four umbrellas ie. Research and Development, Technologies and practices, Infrastructure and Capacity building. To meet the adaptation and mitigation needs in long term, the relevant ongoing schemes(a few are mentioned in the table.1) would be up scaled and a few new interventions(table.2) can also be necessary to include in the PoA .

Table 16.2 Suggestive areas for preparation of projects under NMSA

Focus area	Thrust areas
Research and Development	Customization of hybrid or high yielding varieties of seeds to the specific needs of each Agro-Climatic Zone (ACZ)
	Develop drought and pest resistant crop varieties
	Development of crops with enhanced water and nitrogen use efficiency
	Gene manipulation for introducing C4 pathway in important C3 crops(higher carbon pathways in lower carbon plants)
	Gene manipulation to improve drought tolerant varieties
	Research on micro irrigation to paddy crop
	Research on Bio remediation of polluted lands
	Develop drought tolerant cattle breeds as well as other small ruminants
	Develop better fodder varieties
	Use of Biotechnology to develop climate resilient plant varieties and animal breeds

Table 16.2 Contd...

Focus area	Thrust areas
	Development of localized (mandal/Block) long range weather forecast*
	Development of predictive models for pest and disease surveillance
	Development of modified crop calendars for each agro climatic zones in the changing monsoon scenario*
	Research on Market requirements, new insurance products, credit assessments
	Citizen centric technologies to detect ground water status at farm level*
Technology and practices	customized technologies and packages of practices that are location specific
	Promotion of existing technologies and practices for natural resource conservation
	Propagation of Technologies and practices that increase the mitigation potential at the farm level
	Incentivize the use of local plant varieties and local breeds of cattle*
	Promote Dry seeding of paddy*
	Demarcate and conserve the native organic zones like hilly tracts*
	Promotion of In situ Soil and moisture Conservation measures
	Change in dietary practices of livestock to curb methane emissions from enteric fermentation
	Incentivize lifting of farm residues to avoid in situ burning*
	Recycling of wastes and their conversion into easily transportable and usable forms for their effective utilization in plant nutrient supply
	Soil mapping for microbial population*
	Ground water mapping to the scale of village*
	Link fertilizer subsidies with soil fertility and extent of holding*
	Using technology in crop area and yield estimation as well as crop loss estimation during natural calamities*
	Preparation of water and carbon budgets at village level with people's participation*

Table 16.2 Contd...

Focus area	Thrust areas
Infrastructure Development	Infrastructure development in the water and power sectors
	a)Repairs to irrigation channels*
	b)Cleaning of water bodies*
	c)Improvement in Irrigation water use Efficiency*
	d)Promotion of renewable energy sources
	Infrastructure development to improve rural connectivity
	Promotion of village level grain and seed storage structures
	Promotion of low cost hydroponics to grow green fodder*
	Development of Fodder and Food Banks with the help of Self-help groups
	Establishment of rain gages at revenue village level*
	Establishment of mini weather stations at disaggregated level*
	Rescheduling of crop/ livestock loans at zero percent interest in the event of natural calamities like drought, floods etc.*
Capacity Building	Expand scope to impart training to farmers and other stakeholders including officials of agriculture and allied sectors, irrigation, rural development and environment departments on climate resilient technologies and practices*
	Promotion of demonstrations for the region specific practices and technologies for a larger array of crops
	Development of unified structure for training of farmers and officials
	Use of ICT including mobile technology at farmer doorsteps
	Establishment of Kiosks at village level to be maintained through facilitators*
	Establishment of Village Resource Centres for Agrl Intelligence*

*Those interventions are not mentioned in NMSA Mission document.
These are a few suggestive interventions

D. General guidelines to the extension functionaries or implementing agencies to implement a project.

> (a) *Chalk down the project details:* By taking the confidence of all the stakeholders, and by keeping the aims and objectives of NMSA, one should plan an appropriate project. Problem analysis has to be done to assess the need of that intervention before taking up the project. Define the scope of the project distinctly and precisely. Develop the project plan and verify that the goals of the key elements are clearly defined and closely aligned. It is necessary to establish measurable and trackable success criteria ic Key Performance Indicators, including accomplishing tasks on schedule, achieving budget targets, confirming product functionality or service delivery is satisfactory to the farmer, under the regulatory control of the outlines designed under NMSA.

> (b) *Identify key milestones:* It is necessary to identify key milestones throughout the project. It is necessary to provide a life cycle of the project by including the four main phases: initiation, planning, execution, and closure. It is always better to perform a real evaluation at the end of each phase. Fixing the milestones would provide a platform to make course corrections and revitalize ourselves.

> (c) *Make the communication channels open:* One of the most important steps in project management is to ensure communication lines are open. Throughout the project all the stakeholders should be in touch with each other for which the implementing agency should facilitate a consistent, transparent communication channel. Ensure that everyone has the information that is necessary to make decisions and proceed with the project.

> (d) *Attain Pertinent Documentation:* From project planning to closure the relevant documentation has to be attained throughout the project cycle by covering all the milestones. This would help transparency and to convince all the stakeholders. If it is a successful project the documentation would help to replicate and upscale it.

Test Deliverables: At the end of each milestone the deliverables are to be tested for its quality and ensure that those are in alignment with the project plan.

Evaluate the project: During mid course and at the end of the project it should be evaluated. Evaluation would help to minimize failures and maximize success in the future projects.

E. **Mission Implementation:**   At present NMSA interventions are being implemented through four programmes. Those are Rainfed Area Development (RAD), On Farm Water Management (OFWM), Soil Health Management (SHM), and Climate Change and Sustainable Agriculture: Monitoring, Modeling and Networking (CCSAMMN).There are four stages in implementation of the proposed plans or projects under NMSA viz. 1)Preparation of Mission Implementation Plan(MIP), 2)Preparation of Annual Action Plan (AAP), 3) Project submission and approval and finally 4)Project implementation.

   (i) *Mission Implementation Plan(MIP):* The Mission Implementation Plan(MIP) has to be submitted by the States to Centre. These MIPs should indicate the Action Plans(AAP) and strategies which are culled from District and state plans. These MIPs are to prepare with a holistic view of sustainable agriculture with a climate resilient approach for a horizon of 5 years. These MIPs should be prepared by considering agro climatic characteristics of that geography which are in general need to consider the recommendations made for agriculture sector in State   Action   Plan   on   Climate Change(SAPCC).(Tentative contents of a MIP are shown in Annexure-i)

   (ii) *Annual Action Plan (AAP):* Guiding by the tentative outlay provided for each component by the centre except for Soil Health Management (SHM), the states have to prepare Annual Action Plan (AAP) to operationalize MIP. The designated nodal agency for the state has to coordinate with all implementing departments /agencies and collate the proposals. Care should be taken while preparing the AAPs to integrate District Agriculture Plans and State Agriculture Plans and those activities including RKVY should not be overlapped and they need to be complemented each other. While district implementing agencies are preparing Annual Work Plans, the states have to prepare AAP by consolidating all the AWP. By adopting bottom up approach and taking into confidence of all stakeholders and based on their priorities, the AWPs are to be prepared.

(iii) *Project submission and approval:* By the end of March the sates have to submit separate proposals for each component for DAC's approval.

(iv) *Programme Implementation:* At state level ,State Agriculture department would act as nodal agency, and there is a flexibility of the sates to deploy implementing agency at district level. Different levels of PanchayatRaj institutions should be taken into consideration while preparing and implementing the plans. The location specific interventions are needs to be prepared based on District Agriculture contingency plans and NICRA findings.

F. Financial Implications: The proposed adaptation and mitigation strategies would require an additional budgetary support of Rs. 1,08,000 crors at current prices upto the end of XII plan . A major portion (60% i.e. Rs. 65,000 Crores) would be utilized to adopt technology solutions for mitigating risks related to climate change. Infrastructure development and R&D together will be allocated 35% (Rs. 31,500 Crores)of the total resources, whereas about 5% of allocation(Rs. 5,000 Crores) will be deployed for capacity building.

Since this NMSA would work in a multi-functional approach to obtain an effective responses for mitigating climate change related risks in order to transform our agricultural production system into a resilient mode, it demands a multi-tier institutional mechanism. Such framework would ensure coordination, cooperation and collaboration between Ministries, Departments and stakeholders to achieve the desired objectives. A three-tier institutional mechanism (District, State, Country level) for ensuring convergence at all levels with other Missions is an essential pre-requisite under the National Mission for Sustainable Agriculture (NMSA).

References

NMSA National Mission for Sustainable Agriculture (NMSA) Mission Document, Department of Agriculture and Cooperation, Ministry of Agriculture, GoI, 2010.

Climate Change Perspectives from India, UNDP, 2009 http://www.villanovau.com/resources/project-management/project-management-tips

National Action Plan on Climate Change (NAPCC), Prime Minister's Council on Climate Change, Govt. of India, New Delhi, India, 2008

National Mission for Sustainable Agriculture(NMSA) Operational Guidelines, Department of Agriculture, and Cooperation, Ministry of Agriculture, GoI, 2014

Appendix - I

National Mission for Sustainable Agriculture (NMSA) Mission Implementation Plan (MIP)

1. Introduction:
   (a) About the State
   (b) Agro-climatic features including mean annual rainfall, soil type/depth etc.
   (c) Land use d. Extent of degradation/problem soil
   (d) Available water resources
   (e) Poverty and social development issues
   (f) Strength, Weakness, Opportunity and Challenges (SWOC) Analysis based on PRA exercise
2. The Plan of action for XII Plan:
   (a) Description
   (b) Site selection criteria and process
   (c) Approach and strategy
   (d) Institution
3. Mission Components:
   (a) Rainfed Area Development
       (i) Programme area, Interventions
   (b) Soil Health Management
       (i) Programme area, Interventions
   (c) On Farm Water management
       (i) Programme area, Interventions
   (d) Research/pilot/networking projects on Climate Change
       (i) No and type of projects
4. Mission Management:
   (a) Organization Set-up

    (b) Implementation Arrangements

    (c) Capacity Building of the Staff

    (d) Partnering Institutions

    (e) Selection of Support Organizations

    (f) Monitoring & Evaluation

    (g) IT Infrastructure Development

    (h) Governance & Accountability

5. Community Participation:

    (a) Project cycle at the community level

    (b) Community institutions to be formed/facilitated

    (c) Role of community groups and institutions in the planning and implementation process

    (d) Content and details of community level plans

    (e) Community level implementation arrangements

6. Mission cost

7. Mission outcome & sustainability

Group Activity for Participants

Problem Tree Analysis –
Know the World around You

K. Padmaja

Centre for Good Governance, Hyderabad.

What it is, why it is and how it is

To take up any project or to do interventions in field it is necessary to the implementing agency or to the executing body to analyze the problems of the geography. As the Climate Smart Agriculture interventions are very location specific it is essential to do the problem analysis for every particular location. The Problem Tree method is a method which will preferably conducted in a participatory mode by involving all stakeholders. This method is a planning method based on the needs of a particular location. While going through the process, taking the different steps, there is a continuous space for opportunities, new ideas and contributions from the stakeholders involved. This method will address the real needs of the beneficiaries. It would result in an image of reality which will be useful in formulation of any project or to take up any intervention with the objectives which are accepted and supported by the stakeholders involved in the process.

A properly planned project addressing the real needs of the beneficiaries is necessarily based upon a correct and complete analysis of the existing situation. The existing situation should be interpreted according to the views, needs, interests and activities of parties concerned. It is essential that all those involved accept the plans and the activities or interventions proposed are to be realistic and the stakeholders involved are committed to implement.

There are three steps involved in this analysis, the first one would be the problem analysis which has three components viz. PROBLEM, CAUSE and EFFECT. Once the parties involved identify the problem, they should search for causes and the consequent effects. The second step would be developing objectives with an image of future for better situation. The third step would be analysis of strategies/options available to fulfill the objective. Once the participants come to a common understanding about the problem, cause and effect, the analysis should be presented in the form of diagram or a tree. Depicting the problem in tree form will visualize the real situation to the beneficiaries. After knowing the root cause of the problem the participants can transform these causes into the objectives to encounter the problem. But care should be taken that the objectives should be realistic and practically feasible ones. For example one of the root cause for non adoption of seed treatment with rhizobium culture in legumes in a particular location is non availability of viable rhizobium inoculum. Based on the root cause identified through the problem tree analysis it would be easy to the planners to transform the negativity of the root cause into a positive form to frame the objective ie make the inoculum available to the farmers within the vicinity of the particular village. Once the objective is framed, the implementing agency could develop strategies to achieve the objective.

The purpose of this group activity is to understand the causes related directly or indirectly to Climate Change and the consequent effects.

Duration: 30 minutes

Materials: Pencil, Chart

Process

1. The problem that is assigned to the Group i.e Climate Change is the group's starter problem. Make this problem as the trunk of a tree.

2. With group participation brain storm the cause for the problem and represent those causes as the roots of the tree (problem)

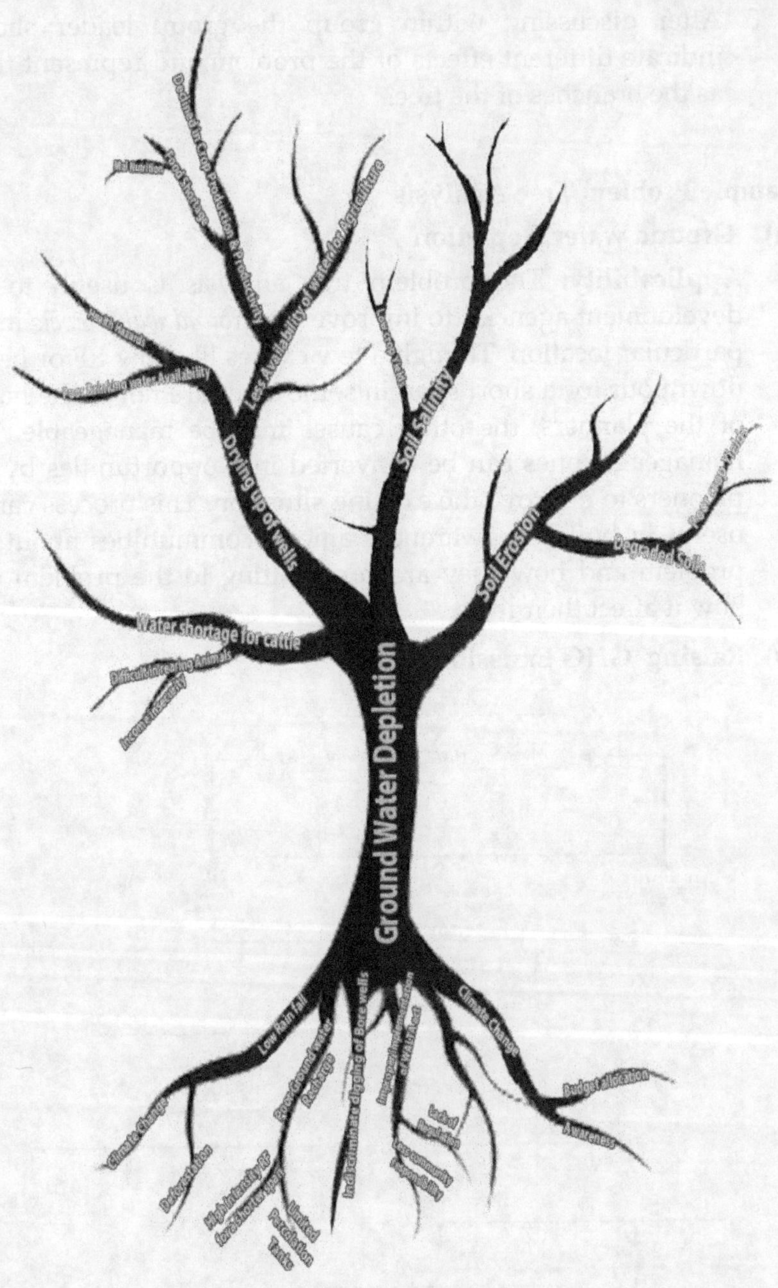

Mal Nutrition
Food Shortage
Declined in Crop Production & Productivity
Less Availability of water for Agriculture
Health Hazard
Poor Drinking water Availability
Less Availability
Less crop yield
Drying up of wells
Soil Salinity
Soil Erosion
Degraded Soils
Poor crop yield
Water shortage for cattle
Difficulty in rearing Animals
Income Insecurity
Ground Water Depletion
Low Rain fall
Over Ground water Recharge
Indiscriminate digging of Bore wells
Improper implementation of Watershed
Climate Change
Deforestation
High intensity of forest shorter crops
Limited Percolation Tanks
Lack of Regulation
Less community participation
Climate Change
Budget allocation
Awareness

3. After discussing within group the group leader should indicate different effects of the problem and represent them as the branches of the tree.

Sample Problem Tree Analysis

(a) Ground water Depletion

Applicability: The problem tree analysis is useful to the development agencies to improve the *ground water levels* in the particular location. Though a few causes like low RF or heavy downpour for a short span in some cases are not in the hands of the planners, the other causes may be manageable. The manageable ones can be converted into opportunities by the planners to improve the existing situation. This process can be useful in building awareness among communities about the problem and how they are contributing to the problem and how it affect their lives.

(b) Raising GHG Emissions

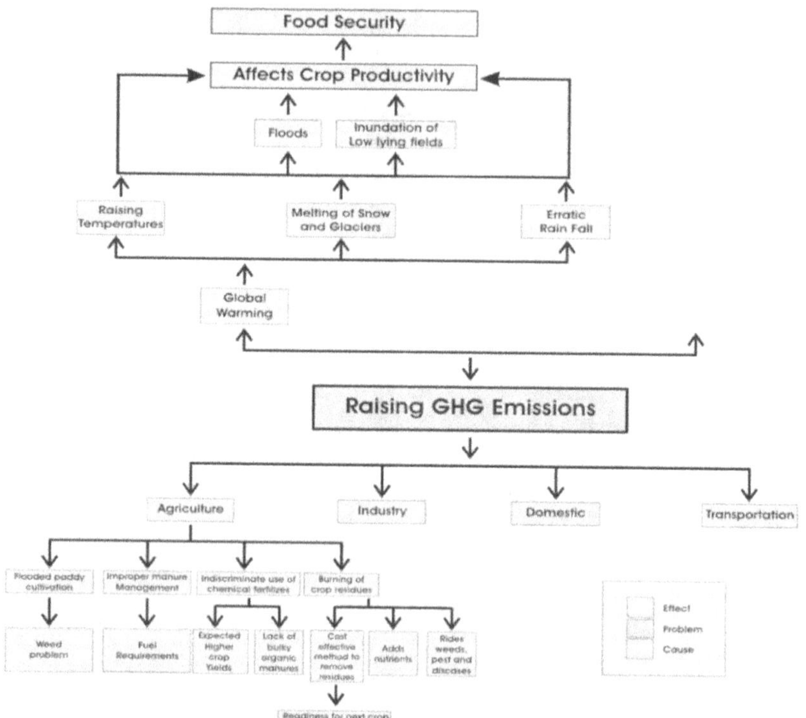

References

European Commission (Editor) (2004): Aid Delivery Methods, Project Cycle Management Guidelines. Brussels: European Commission - EuropeAid Cooperation Office.

MDF (Editor), 2005: MDF Tool: Problem Tree Analysis: MDF Training and Consultancy.

Glossary of Common Climate Change Terms

Adaptation: ways of coping with the impacts of global warming on species, ecosystems and human society.

Anthropogenic: caused by human activities.

Aerosols: Solid or liquid particles suspended within the atmosphere (see "sulfate aerosols" and "black carbon aerosols").

Afforestation: Planting of new forests on lands that have not been recently forested.

Adaptive Capacity: Adaptive Capacity is the ability of a system to adjust to climate change (including climate variability and extremes) to moderate potential damages, to take advantage of opportunities, or to cope with the consequences. Also see the page explaining the link between adaptation and mitigation.

Additionality: The Kyoto Protocol articles on Joint Implementation (Art. 6) and the Clean Development Mechanism (Art. 12) state that emissions reduction units (ERUs and CERs) will be awarded to project-based activities provided that the projects achieve emissions reductions that are 'additional to those that otherwise would occur.'

Agroecology: Agroecology is the science of sustainable agriculture; the methods of agroecology have as their goal achieving sustainability of agricultural systems balanced in all spheres. This includes the socio-economic and the ecological or environmental.

Albedo: Refers to the ratio of light from the sun that is reflected by the Earth's surface to the light received by it. Unreflected light is converted to infrared radiation (i.e., heat), which causes atmospheric warming (see "radiative forcing"). Thus, surfaces with a high albedo (e.g., snow and ice) generally contribute to cooling, whereas surfaces with a low albedo (e.g., forests) generally contribute to warming. Changes in land use that significantly alter the characteristics of land surfaces can therefore influence the climate through changes in albedo.

Allocation: Under an emissions trading scheme, permits to emit can initially either be given away for free, usually under a 'grandfathering' approach based on past emissions in a base year or an 'updating' approach based on the more recent emissions. The alternative is to auction permits in an initial market offering.

Ancillary Benefits: Complementary benefits of a climate policy including improvements in local air quality and reduced reliance of imported fossil fuels.

Anthropogenic Emissions: Emissions of greenhouse gasses resulting from human activities.

Assigned Amount: In the Kyoto Protocol, the permitted emissions, in CO_2 equivalents, during a commitment period. It is calculated using the Quantified Emission Limitation and Reduction Commitment (QELRC), together with rules specifying how and what emissions are to be counted.

Biodiversity: life in all its forms, essential to maintaining functioning ecosystems that provide services essential for human survival and quality of life.

Biomass: The total dry organic matter or stored energy content of living organisms. Biomass can be used for fuel directly by burning it (e.g. wood), indirectly by fermentation to an alcohol (e.g. sugar) or extraction of combustible oils (e.g. soybeans).

Base Year: Targets for reducing GHG emissions are often defined in relation to a base year. In the Kyoto Protocol, 1990 is the base year for most countries for the major GHGs; 1995 can be used as the base year for some of the minor GHGs.

Baselines: The baseline estimates of population, GDP, energy use and hence resultant greenhouse gas emissions without climate policies, determine how big a reduction is required, and also what the impacts of climate change without policy will be.

Basket of Gases: This refers to the group six of greenhouse gases regulated under the Kyoto Protocol. They are listed in Annex A of the Kyoto Protocol and include: carbon dioxide (CO_2), methane

(CH₄), nitrous oxide (N₂O), hydrofluorocarbons (HFCs), perfluorocarbons (PFCs), and sulphur hexafluoride (SF₆).

Berlin Mandate: Decision of the Parties reached at the first session of the Conference of the Parties to the UNFCCC (COP-1) in 1995 in Berlin that the commitments made by Annex I countries were inadequate and thus needed to be strengthened.

Biodiversity: The variety of organisms found within a specified geographic region.

Black Carbon Aerosols: Particles of carbon in the atmosphere produced by inefficient combustion of fossil fuels or biomass. Black carbon aerosols absorb light from the sun, shading and cooling the Earth's surface, but contribute to significant warming of the atmosphere (see "radiative forcing").

Bryd-Hagel Resolution: In June 1997, anticipating the December 1997 meeting in Kyoto, Senator Robert C. Byrd (D-WV) introduced, with Sen. Chuck Hagel (R-NE) and 44 other cosponsors, a resolution stating that the impending Kyoto Protocol (or any subsequent international climate change agreement) should not - "(A) mandate new commitments to limit or reduce GHG emissions for the Annex I Parties [i.e. industrialized countries], unless the protocol or other agreement also mandates new specific scheduled commitments to limit or reduce GHG emissions for Developing Country Parties within the same compliance period, or (B) would result in serious harm to the economy of the United States..."

Bubble: An option in the Kyoto Protocol that allows a group of countries to meet their targets jointly by aggregating their total emissions. The member states of the European Union are utilizing this option.

Carbon Capture and Storage (CCS): the process of capturing greenhouse gas pollution from coal or gas power plants and storing it underground instead of releasing it into the atmosphere. CCS could reduce emissions from a power plant by up to 80%.

Carbon Credit: used in emissions trading schemes (see emissions trading), where one credit gives the owner the right to emit one tonne of CO₂.

Carbon Cycle: the exchange and movement of carbon – a chemical element – through the atmosphere, oceans and water, living things, soils and geological deposits.

Carbon Neutral: where an individual or company's carbon emissions are effectively reduced to zero through a combination of reducing energy consumption, using renewable energy and offsetting the remainder by (for example) planting trees to absorb carbon dioxide from the atmosphere.

Carbon Offsetting: where an investment is made in a project that will lead to the prevention or removal of carbon dioxide from the atmosphere (for example, planting trees or building renewable energy power stations to avoid the construction of coal ones).

Carbon Price: puts a price on greenhouse gas emissions to create a disincentive for their release (and incentive to capture or avoid them). A carbon price can be imposed through a carbon tax, an emissions trading scheme (which fixes the emission level and allows price to vary) or a variety of other mechanisms.

Carbon Sequestration: the uptake and storage of carbon (e.g., by trees, or burying underground).

Clean Coal: technologies designed to reduce the amount of CO_2 emitted by coal-fired power stations.

Climate Change: significant changes from one climatic condition to another, commonly referring to the increase in Earth's surface temperature caused by human activities. Also often called global warming, anthropogenic climate change, anthropogenic global warming and the enhanced greenhouse effect.

Carbon Dioxide (CO_2): CO_2 is a colorless, odorless, non-poisonous gas that is a normal part of the ambient air. Of the six greenhouse gases normally targeted, CO_2 contributes the most to human-induced global warming. Human activities such as fossil fuel combustion and deforestation have increased atmospheric concentrations of CO_2 by approximately 30 percent since the industrial revolution. CO_2 is the standard used to determine the "global warming potentials" (GWPs) of other gases. CO_2 has been assigned a 100-year GWP of 1 (i.e., the warming effects over a 100-year time frame relative to other gases).

Carbon Dioxide Equivalent (CO₂e): Carbon Dioxide Equivalent (CO₂e). The emissions of a gas, by weight, multiplied by its "global warming potential."

Carbon Sinks: Processes that remove more carbon dioxide from the atmosphere than they release. Both the terrestrial biosphere and oceans can act as carbon sinks.

Carbon Taxes: A surcharge on the carbon content of oil, coal, and gas that discourages the use of fossil fuels and aims to reduce carbon dioxide emissions.

Certified Emissions Reduction (CER): Reductions of greenhouse gases achieved by a Clean Development Mechanism (CDM) project. A CER can be sold or counted toward Annex I countries' emissions commitments. Reductions must be additional to any that would otherwise occur.

Chlorofluorocarbons (CFCs): CFCs are synthetic industrial gases composed of chlorine, fluorine, and carbon. They have been used as refrigerants, aerosol propellants, cleaning solvents and in the manufacture of plastic foam. There are no natural sources of CFCs. CFCs have an atmospheric lifetime of decades to centuries, and they have 100-year "global warming potentials" thousands of times that of CO₂, depending on the gas. In addition to being greenhouse gases, CFCs also contribute to ozone depletion in the stratosphere and are controlled under the Montreal Protocol.

Clean Development Mechanism (CDM): One of the three market mechanisms established by the Kyoto Protocol. The CDM is designed to promote sustainable development in developing countries and assist Annex I Parties in meeting their greenhouse gas emissions reduction commitments. It enables industrialized countries to invest in emission reduction projects in developing countries and to receive credits for reductions achieved.

Climate: The long-term average weather of a region including typical weather patterns, the frequency and intensity of storms, cold spells, and heat waves. Climate is not the same as weather.

Climate Change: Refers to changes in long-term trends in the average climate, such as changes in average temperatures. In IPCC usage, climate change refers to any change in climate over time, whether due to natural variability or as a result of human activity.

In UNFCC usage, climate change refers to a change in climate that is attributable directly or indirectly to human activity that alters atmospheric composition.

Climate Sensitivity: The average global air surface temperature change resulting from a doubling of pre-industrial atmospheric CO_2 concentrations. The IPCC estimates climate sensitivity at 1.5-4.5 ºC (2.7-8.1ºF).

Climate Variability: Refers to changes in patterns, such as precipitation patterns, in the weather and climate.

Commitment Period: The period under the Kyoto Protocol during which Annex I Parties' GHG emissions, averaged over the period, must be within their emission targets. The first commitment period runs from January 1, 2008 to December 31, 2012.

Conference of the Parties (COP): The supreme decision-making body comprised of the parties that have ratified the UN Framework Convention on Climate Change. It meets on an annual basis. As of February 2003, it is comprised of 188 countries.

Dangerous Climate Change: a level of climate change that will have severe impacts on societies, economies and the natural world. WWF defines dangerous climate change as a rise in average global surface temperatures of 2 °C or more (above pre-Industrial Revolution average surface temperatures).

Discounting: The process that reduces future costs and benefits to reflect the time value of money and the common preference of consumption now rather than later.

Early Crediting: A provision that allows crediting of emission reductions achieved prior to the start of a legally imposed emission control period. These credits can then be used to assist in achieving compliance once a legally imposed system begins.

Ecosystem: A community of organisms and its physical environment.

Emissions: The release of substances (e.g., greenhouse gases) into the atmosphere.

Energy Efficiency: the amount of energy needed to provide the same amount of heating, cooling or other energy service from different sources, appliances or systems.

Emissions Cap: A mandated restraint in a scheduled timeframe that puts a "ceiling" on the total amount of anthropogenic greenhouse gas emissions that can be released into the atmosphere. This can be measured as gross emissions or as net emissions (emissions minus gases that are sequestered).

Emissions Reduction Unit (ERU): Emissions reductions generated by projects in Annex B countries that can be used by another Annex B country to help meet its commitments under the Kyoto Protocol. Reductions must be additional to those that would otherwise occur.

Emissions Trading: A market mechanism that allows emitters (countries, companies or facilities) to buy emissions from or sell emissions to other emitters. Emissions trading is expected to bring down the costs of meeting emission targets by allowing those who can achieve reductions less expensively to sell excess reductions (e.g. reductions in excess of those required under some regulation) to those for whom achieving reductions is more costly.

Energy Resources: The available supply and price of fossil and alternative resources will play a huge role in estimating how much a greenhouse gas constraint will cost. In the U.S. context, natural gas supply (and thus price) is particularly important, as it is expected to be a transition fuel to a lower carbon economy.

Enhanced Greenhouse Effect: The increase in the natural greenhouse effect resulting from increases in atmospheric concentrations of GHGs due to emissions from human activities.

Entry Into Force: The point at which international climate change agreements become binding. The United Nations Framework Convention on Climate Change (UNFCCC) has entered into force. In order for the Kyoto Protocol to do so as well, 55 Parties to the Convention must ratify (approve, accept, or accede to) the Protocol, including Annex I Parties accounting for 55 percent of that group's carbon dioxide emissions in 1990. As of June 2003, 110 countries had ratified the Protocol, representing 43.9 percent of Annex I emissions.

Evapotranspiration: The process by which water re-enters the atmosphere through evaporation from the ground and transpiration by plants.

El Niño: Translates from Spanish as 'the boy-child'. Peruvian fisherman originally used the term referring to the Christ as a child. This was used to describe the appearance, around Christmas, of a warm ocean current off the South American coast. Today, the term El Niño refers to the extensive warming of the central and eastern Pacific that leads to a major shift in weather patterns across the Pacific. In Australia, and more so for eastern Australia, El Niño events are associated with an increased probability of drier conditions.

Extratropical Cyclones: Sometimes called mid-latitude cyclones, are a group of cyclones defined as synoptic scale low pressure weather systems that occur in the middle latitudes of the Earth having neither tropical nor polar characteristics.

Fossil Fuel: Fuel of biological (plant and animal) origin that has become fossilised over millions of years. Largely comprised of carbon and hydrogen, coal, natural gas and oil are all fossil fuels.

GDP: Gross Domestic Product, a measure of overall economic activity.

Greenhouse Gases: gases such as carbon dioxide, water vapour, methane, nitrous oxide, ozone, and various fluorocarbons that absorb and re-emit heat from the solar radiation that hits the Earth into the atmosphere.

Greenhouse Effect: the effect created by the band of greenhouse gases that blanket the Earth. The greenhouse effect keeps the Earth's surface within a temperature range that makes life on Earth (as we know it) possible.

Greenhouse Pollution: pollution by humans of the Earth's atmosphere by the release of excessive greenhouse gases. This increases the volume of gases in the atmosphere, traps more solar radiation and leads to global warming.

Greenhouse Gas Intensity: the volume of greenhouse gases emitted per unit of energy or economic output. It is a relative measure in that, if the economy is growing, greenhouse intensity per unit of economic output may be falling but greenhouse gas emissions may be increasing in absolute terms. Greenhouse gas intensity is to be contrasted with greenhouse gas emission reductions, where the volume of gases emitted falls in absolute terms.

General Circulation Model (GCM): A computer model of the basic dynamics and physics of the components of the global climate system (including the atmosphere and oceans) and their interactions which can be used to simulate climate variability and change.

Global Warming: The progressive gradual rise of the Earth's average surface temperature thought to be caused in part by increased concentrations of GHGs in the atmosphere.

Global Warming Potential (GWP): A system of multipliers devised to enable warming effects of different gases to be compared. The cumulative warming effect, over a specified time period, of an emission of a mass unit of CO_2 is assigned the value of 1. Effects of emissions of a mass unit of non-CO_2 greenhouse gases are estimated as multiples. For example, over the next 100 years, a gram of methane (CH_4) in the atmosphere is currently estimated as having 23 times the warming effect as a gram of carbon dioxide; methane's 100-year GWP is thus 23. Estimates of GWP vary depending on the time-scale considered (e.g., 20-, 50-, or 100-year GWP), because the effects of some GHGs are more persistent than others.

Greenhouse Effect: The insulating effect of atmospheric greenhouse gases (e.g., water vapor, carbon dioxide, methane, etc.) that keeps the Earth's temperature about 60"F warmer than it would be otherwise.

Greenhouse Gas (GHG): Any gas that contributes to the "greenhouse effect."

Group of 77 and China, or G77/China: An international organization established in 1964 by 77 developing countries; membership has now increased to 133 countries. The group acts as a major negotiating bloc on some issues including climate change.

HGWP (High Global Warming Potential): Some industrially produced gases such as sulfur hexafluoride (SF_6), perfluorocarbons (PFCs), and hydrofluorocarbons (HFCs) have extremely high GWPs. Emissions of these gases have a much greater effect on global warming than an equal emission (by weight) of the naturally occurring gases. Most of these gases have GWPs of 1,300 - 23,900 times that of CO_2. These GWPs can be compared to the

GWPs of CO_2, CH_4, and N_2O which are presently estimated to be 1, 23 and 296, respectively.

Hot Air: A situation in which emissions (of a country, sector, company or facility) are well below a target due to the target being above emissions that materialized under the normal course of events (i.e. without deliberate emission reduction efforts). Hot air can result from over-optimistic projections of growth. Emissions are often projected to grow roughly in proportion to GDP, and GDP is often projected to grow at historic rates.

Hydrofluorocarbons (HFCs): HFCs are synthetic industrial gases, primarily used in refrigeration and semi-conductor manufacturing as commercial substitutes for chlorofluorocarbons (CFCs). There are no natural sources of HFCs. The atmospheric lifetime of HFCs is decades to centuries, and they have 100-year "global warming potentials" thousands of times that of CO_2, depending on the gas. HFCs are among the six greenhouse gases to be curbed under the Kyoto Protocol.

Incentive-based Regulation: A regulation that uses the economic behavior of firms and households to attain desired environmental goals. Incentive-based programs involve taxes on emissions or tradable emission permits. The primary strength of incentive-based regulation is the flexibility it provides the polluter to find the least costly way to reduce emissions.

Intergenerational Equity: The fairness of the distribution of the costs and benefits of a policy when costs and benefits are borne by different generations. In the case of a climate change policy the impacts of inaction in the present will be felt in future generations.

Intergovernmental Panel on Climate Change (IPCC): The IPCC was established in 1988 by the World Meteorological Organization and the UN Environment Programme. The IPCC is responsible for providing the scientific and technical foundation for the United Nations Framework Convention on Climate Change (UNFCC), primarily through the publication of periodic assessment reports (see "Second Assessment Report" and "Third Assessment Report").

Joint Implementation (JI): One of the three market mechanisms established by the Kyoto Protocol. Joint Implementation occurs when an Annex B country invests in an emissions reduction or sink enhancement project in another Annex B country to earn emission reduction units (ERUs).

Kyoto Mechanisms: The Kyoto Protocol creates three market-based mechanisms that have the potential to help countries reduce the cost of meeting their emissions reduction targets. These mechanisms are Joint Implementation (Article 6), the Clean Development Mechanisms (Article 17).

Kyoto Protocol: An international agreement adopted in December 1997 in Kyoto, Japan. The Protocol sets binding emission targets for developed countries that would reduce their emissions on average 5.2 percent below 1990 levels.

Land Use, Land-Use Change and Forestry (LULUCF): Land uses and land-use changes can act either as sinks or as emission sources. It is estimated that approximately one-fifth of global emissions result from LULUCF activities. The Kyoto Protocol allows Parties to receive emissions credit for certain LULUCF activities that reduce net emissions.

Mitigation of Global Warming: actions to reduce or avoid greenhouse gas emissions, in order to avoid global warming.

MRET (Mandatory Renewable Energy Target): a scheme to increase the amount of electricity generated from renewable energy sources.

Market Benefits: Benefits of a climate policy that can be measured in terms of avoided market impacts such as changes in resource productivity (e.g., lower agricultural yields, scarcer water resources) and damages to human-built environment (e.g., coastal flooding due to sea-level rise).

Mauna Loa Record: The record of measurement of atmospheric CO_2 concentrations taken at Mauna Loa Observatory, Mauna Loa, Hawaii, since March 1958. This record shows the continuing increase in average annual atmospheric CO_2 concentrations.

Methane: The simplest hydrocarbon, methane, is a gas (at standard temperature and pressure) with a chemical formula of CH_4. However, when averaged over 100 years each kg of CH_4 warms the earth 23 times as much as the same mass of CO_2.

CH_4 is among the six greenhouse gases to be curbed under the Kyoto Protocol. Atmospheric CH_4 is produced by natural processes, but there are also substantial emissions from human activities such as landfills, livestock and livestock wastes, natural gas and petroleum systems, coalmines, rice fields, and wastewater treatment. CH_4 has a relatively short atmospheric lifetime of approximately 10 years, but its 100-year GWP is currently estimated to be approximately 23 times that of CO_2.

Mitigation: In the context of climate change is any action taken to permanently eliminate or reduce the long-term risk to human life, property, and function from the hazards of climate change. Also see the page explaining the relationship between adaptation and mitigation.

Microwave Sounding Units (MSU): Sensors carried aboard Earth orbiting satellites that have been used since 1979 to monitor tropospheric temperatures.

Montreal Protocol: (on Substances that Deplete the Ozone Layer) An international agreement that entered into force in January 1989 to phase out the use of ozone-depleting compounds such as methyl chloroform, carbon tetrachloride, and CFCs. CFCs are potent greenhouse gases which are not regulated by the Kyoto Protocol since they are covered by the Montreal Protocol.

National Action Plans: Plans submitted to the Conference of the Parties (COP) by all Parties outlining the steps that they have adopted to limit their anthropogenic GHG emissions. Countries must submit these plans as a condition of participating in the UN Framework Convention on Climate Change and, subsequently, must communicate their progress to the COP regularly.

Negative Feedback: A process that results in a reduction in the response of a system to an external influence. For example, increased plant productivity in response to global warming would be a negative feedback on warming, because the additional growth would act as a sink CO_2, reducing the atmospheric CO_2 concentration.

Nitrous Oxide (N_2O): N_2O is among the six greenhouse gases to be curbed under the Kyoto Protocol. N_2O is produced by natural processes, but there are also substantial emissions from human activities such as agriculture and fossil fuel combustion. The atmospheric lifetime of N_2O is approximately 100 years, and its 100-year GWP is currently estimated to be 296 times that of CO_2.

Non-Annex B Parties: Countries that are not listed in Annex B of the Kyoto Protocol.

Non-Annex I Parties: Countries that have ratified or acceded to the UNFCCC that are not listed in Annex I of the UNFCCC.

Non-Market Benefits: Benefits of a climate policy that can be measured in terms of avoided non-market impacts such as human-health impacts (e.g., increased incidence of tropical diseases) and damages to ecosystems (e.g., loss of biodiversity).

Non-Party: A state that has not ratified the UNFCCC. Non-parties may attend talks as observers.

Ozone: About 90% of the ozone in our atmosphere is contained in the stratosphere, the region from about 10 to 50 km (32,000 to 164,000 feet) above Earth's surface. Although the concentration of ozone in the ozone layer is very small, it is vitally important to life because it absorbs biologically harmful ultraviolet (UV) radiation emitted from the Sun. Some breakdown of the Ozone layer can be expected to continue due to CFCs used by nations which have not banned them, and due to gases which are already in the stratosphere. CFCs have very long atmospheric lifetimes, ranging from 50 to over 100 years, so the final recovery of the ozone layer is expected to require several lifetimes.

Perfluorocarbons (PFCs): PFCs are among the six types of greenhouse gases to be curbed under the Kyoto Protocol. PFCs are synthetic industrial gases generated as a by-product of aluminum smelting and uranium enrichment. They also are used as substitutes for CFCs in the manufacture of semiconductors. There are no natural sources of PFCs. PFCs have atmospheric lifetimes of thousands to tens of thousands of years and 100-year GWPs thousands of times that of CO_2, depending on the gas.

Ppm or ppb: parts per million or parts per billion, a measure of the amount of greenhouse gas emissions in the atmosphere.

Polluter Pays Principle (PPP): The principle that countries should in some way compensate others for the effects of pollution that they (or their citizens) generate or have generated.

Positive Feedback: A process that results in an amplification of the response of a system to an external influence. For example, increased atmospheric water vapor in response to global warming would be a positive feedback on warming, because water vapor is a GHG.

Quantified Emission Limitation and Reduction Commitment (QELRC): Also known as QELRO (Quantified Emission Limitation and Reduction Objective): The quantified commitments for GHG emissions listed in Annex B of the Kyoto Protocol. QELRCs are specified in percentages relative to 1990 emissions.

Renewable Energy: energy derived from the wind, sun, tides and other sources that, for all practical purposes, cannot be depleted.

Radiative Forcing: The term radiative forcing refers to changes in the energy balance of the earth-atmosphere system in response to a change in factors such as greenhouse gases, land-use change, or solar radiation. The climate system inherently attempts to balance incoming (e.g., light) and outgoing (e.g. heat) radiation. Positive radiative forcings increase the temperature of the lower atmosphere, which in turn increases temperatures at the Earth's surface. Negative radiative forcings cool the lower atmosphere. Radiative forcing is most commonly measured in units of watts per square meter (W/m2).

Radiosondes: Sensors carried aboard weather balloons that have been in continuous use since 1979 for the monitoring of tropospheric temperatures.

Ratification: After signing the UNFCCCCor the Kyoto Protocol, a country must ratify it, often with the approval of its parliament or other legislature. In the case of the Kyoto Protocol, a Party must deposit its instrument of ratification with the UN Secretary General in New York.

Reforestation: Replanting of forests on lands that have recently been harvested.

Regional Groups: The five regional groups meet privately to discuss issues and nominate bureau members and other officials. They are Africa, Asia, Central and Eastern Europe (CEE), Latin America and the Caribbean (GRULAC), and the Western Europe and Others Group (WEOG).

Renewable Energy: Energy obtained from sources such as geothermal, wind, photovoltaic, solar, and biomass.

Revenue Recycling: If permits are auctioned, this gives considerable sums of money to be recycled back into the economy, either through a lump sum payment of offsetting other taxes. If the existing taxes that are correspondingly reduced were very inefficient, this allows this allows the possibility of both environmental and economic benefits from the trading system, commonly called the 'double dividend.'

Sea Level Rise: One of the impacts of global warming.

Second Assessment Report (SAR): The Second Assessment Report, prepared by the Intergovernmental Panel on Climate Change, reviewed the existing scientific literature on climate change. Finalized in 1995, it is comprised of three volumes: Science; Impacts, Adaptations and Mitigation; and Economic and Social Dimensions of Climate Change.

Secretariat of the UN Framework Convention: The United Nations staff assigned the responsibility of conducting the affairs of the UNFCCC. In 1996 the Secretariat moved from Geneva, Switzerland, to Bonn, Germany.

Sequestration: Opportunities to remove atmospheric CO_2, either through biological processes (e.g. plants and trees), or geological processes through storage of CO_2 in underground reservoirs.

Sinks: Any process, activity or mechanism that results in the net removal of greenhouse gases, aerosols, or precursors of greenhouse gases from the atmosphere.

Source: Any process or activity that results in the net release of greenhouse gases, aerosols, or precursors of greenhouse gases into the atmosphere.

SRES Scenarios: A suite of emissions scenarios developed by the Intergovernmental Panel on Climate Change in its Special Report on Emissions Scenarios (SRES). These scenarios were developed to explore a range of potential future greenhouse gas emissions pathways over the 21st century and their subsequent implications for global climate change.

Stratosphere: The region of the Earth's atmosphere 10-50 km above the surface of the planet.

Subsidiary Body for Implementation (SBI): A permanent body established by the UNFCCC that makes recommendations to the COP on policy and implementation issues. It is open to participation by all Parties and is composed of government representatives.

Subsidiary Body for Scientific & Tech. Advice (SBSTA): A permanent body established by the UNFCCC that serves as a link between expert information sources such as the IPCC and the COP.

Substitution: The economic process of trading off inputs and consumption due to changes in prices arising from a constraint on greenhouse gas emissions. How the extremely flexible U.S. economy adapts to available substitutes and/or finds new methods of production under a greenhouse gas constraint will be critical in minimizing overall costs of reducing emissions.

Sulfate Aerosols: Sulfur-based particles derived from emissions of sulfur dioxide (SO_2) from the burning of fossil fuels (particularly coal). Sulfate aerosols reflect incoming light from the sun, shading and cooling the Earth's surface (see "radiative forcing") and thus offset some of the warming historically caused by greenhouse gases.

Sulfur Hexafluoride (SF_6): SF_6 is among the six types of greenhouse gases to be curbed under the Kyoto Protocol. SF_6 is a synthetic industrial gas largely used in heavy industry to insulate high-voltage equipment and to assist in the manufacturing of cable-cooling systems. There are no natural sources of SF_6. SF_6 has an atmospheric lifetime of 3,200 years. Its 100-year GWP is currently estimated to be 22,200 times that of CO_2.

Supplementarity: The Protocol does not allow Annex I parties to meet their emission targets entirely through use of emissions trading and the other Kyoto Mechanisms; use of the mechanisms must be supplemental to domestic actions to limit or reduce their emissions.

Targets and Timetables: Targets refer to the emission levels or emission rates set as goals for countries, sectors, companies, or facilities. When these goals are to be reached by specified years, the years at which goals are to be met are referred to as the timetables. In the Kyoto Protocol, a target is the percent reduction from the 1990 emissions baseline that the country has agreed to. On average, developed countries agreed to reduce emissions by 5.2% below 1990 emissions during the period 2008-2012, the first commitment period.

Technological Change: How much technological change will be additionally induced by climate policies is a crucial, but not well quantified, factor in assessing the costs of long-term mitigation of greenhouse gas emissions.

Thermal expansion: Expansion of a substance as a result of the addition of heat. In the context of climate change, thermal expansion of the world's oceans in response to global warming is considered the predominant driver of current and future sea-level rise.

Thermohaline Circulation (THC): A three-dimensional pattern of ocean circulation driven by wind, heat and salinity that is an important component of the ocean-atmosphere climate system. In the Atlantic, winds transport warm tropical surface water northward where it cools, becomes more dense, and sinks into the deep ocean, at which point it reverses direction and migrates back to the tropics, where it eventually warms and returns to the surface. This cycle or "conveyor belt" is a major mechanism for the global transport of heat, and thushas an important influence on the climate. Global warming is projected to increase sea-surface temperatures, which may slow the THC by reducing the sinking of cold water in the North Atlantic. In addition, ocean salinity also influences water density, and thus decreases in sea-surface salinity from the melting of ice caps and glaciers may also slow the THC.

Third Assessment Report (TAR): The most recent Assessment Report prepared by the Intergovernmental Panel on Climate Change, which reviewed the existing scientific literature on climate change, including new information acquired since the completion of the Second Assessment report (SAR). Finalized in 2001, it is comprised of three volumes: Science; Impacts and Adaptation; and Mitigation.

Trace Gas: A term used to refer to gases found in the Earth's atmosphere other than nitrogen, oxygen, argon and water vapor. When this terminology is used, carbon dioxide, methane, and nitrous oxide are classified as trace gases. Although trace gases taken together make up less than one percent of the atmosphere, carbon dioxide, methane and nitrous oxide are important in the climate system. Water vapor also plays an important role in the climate system; its concentrations in the lower atmosphere vary considerably from essentially zero in cold dry air masses to perhaps 4 percent by volume in humid tropical air masses.

Troposphere: The region of the Earth's atmosphere 0-10 km above the planet's surface.

Umbrella Group: Negotiating group within the UNFCCC process comprising the United States, Canada, Japan, Australia, New Zealand, Norway, Iceland, Russia, and Ukraine.

UN Framework Convention on Climate Change (UNFCCC): A treaty signed at the 1992 Earth Summit in Rio de Janeiro that calls for the "stabilization of greenhouse gas concentrations in the atmosphere at a level that would prevent dangerous anthropogenic interference with the climate system." The treaty includes a non-binding call for developed countries to return their emissions to 1990 levels by the year 2000. The treaty took effect in March 1994 upon ratification by more than 50 countries. The United States was the first industrialized nation to ratify the Convention.

Uncertainty: Uncertainty is a prominent feature of the benefits and costs of climate change. Decision makers need to compare risk of premature or unnecessary actions with risk of failing to take actions that subsequently prove to be warranted. This is complicated by potential irreversibilities in climate impacts and long term investments.

Urban Heat Island (UHI): Refers to the tendency for urban areas to have warmer air temperatures than the surrounding rural landscape, due to the low albedo of streets, sidewalks, parking lots, and buildings. These surfaces absorb solar radiation during the day and release it at night, resulting in higher night temperatures.

Vulnerability: Vulnerability is the degree to which a system is susceptible to, or unable to cope with, adverse effects of climate change, including climate variability and extremes. Vulnerability is a function of the character, magnitude, and rate of climate change and variation to which a system is exposed, its sensitivity, and its adaptive capacity.

Vector-borne Disease: Disease that results from an infection transmitted to humans and other animals by blood-feeding anthropods, such as mosquitoes, ticks, and fleas. Examples of vector-borne diseases include Dengue fever, viral encephalitis, Lyme disease, and malaria.

Water Vapor (H_2O): Water vapor is the primary gas responsible for the greenhouse effect. It is believed that increases in temperature caused by anthropogenic emissions of greenhouse gases will increase the amount of water vapor in the atmosphere, resulting in additional warming (see "positive feedback").

Weather: Describes the short-term (i.e., hourly and daily) state of the atmosphere. Weather is not the same as climate. The state of the atmosphere at a point in time. Principal elements of weather include temperature, precipitation, wind, air pressure, and humidity.

About Contributing Authors

Prof. Devi Prasad Juvvadi is currently Director, Agriculture Management at Centre for Good Governance (CGG), Hyderabad. He has about three decades of experience in agricultural research, teaching and consultancy. His basic training was B.Sc (Agric) at APAU, later he pursued Masters in Soil Science at American University of Beirut and advanced training in soil science at Oxford University, England. Prof. Devi Prasad worked as Assistant Professor at American University of Beirut for a decade and adjunct to US-Saudi Joint Commission on Economic Co-operation in Riyadh, Saudi Arabia. He worked as Consultant to various international organizations and projects supported by UNDP, UNECWA, Wold Bank, DFID in Middle East, Africa and India.

After about one and half decades of education and working in various countries, Prof. Devi Prasad returned to India and worked with non profit sector and has been trainer, resource person and consultant to many organizations including DMRL, DRDL, MIDHANI, NAARM, MANAGE, SAMETI, EEI, SBI, ITC, APWELL, AFC etc.

Prof. Devi Prasad worked as Team Leader for World Bank supported project, A.P. Community Based Tank Management Project through Agriculture Finance Corporation of India. He also worked as Team Leader for External M&E of A.P.Well, a Netherlands supported project. For the last decade he is involved in various projects with Centre for Good Governance including change and service delivery projects in agriculture under a DFID supported programme.

Prof Devi Prasad has been member of Soil Science Society of America, American Society of Agronomy, Crop Science Society of America, Saudi Biological Society etc. He has published widely in international journals, published bulletins, review articles, contributed to English and Telugu news papers on agriculture and written policy papers, briefs and few books on agriculture. Prof. Devi Prasad presented papers at several international conferences including World Agriculture Forum, and lectured in various universities in Turkey, Israel and USA. He has travelled widely in over 26 countries in all the continents. His current research interests include agriculture governance, innovation in agriculture, change management etc.

Cherukumalli Srinivasarao, is presently working as Director, CRIDA, Hyderabad.

Dr. Ch. Srinivasa Rao did his Ph.D. from IARI and Post-Doctoral at Tel-Aviv University, Israel. Worked at IISS, Bhopal; IIPR, Kanpur and CRIDA, Hyderabad; Project Coordinator, AICRP for Dryland Agriculture; Deputation to International Crop Research Institute for Semi Arid Tropics (ICRISAT), Patancheru, 2006-2008. Presently working as Director, Central Research Institute for Dryland Agriculture.

His contributions are on participatory soil health management, carbon sequence, climate change, contingency planning and dryland agriculture.

Awards/Honours: Certificate of Merit ISCA, 1995; ISSS Golden Jubilee Young Scientist Award, 1997; ICAR Young Scientist Award, 1998; International Potash Institute-FAI Award, 1998; Pran Vohra Memorial Award, 2000 and Dr. B.C. Deb Memorial Award, 2006 of Indian Science Congress Association; PPIC-FAI Award, 2006; Dhiru Morarjee Memorial Award of FAI, 1993 and 2003; ISPRD Recognition Award, 2006-07; Doreen Mashler Award by ICRISAT, 2007; ICRISAT Millennium Science Award, 2008; International Plant Nutrition Institute Prize 2008; Sukumar Basu Memorial Award of IARI, 2009; Vasantrao Naik Award of ICAR, 2009; FAI Golden Jubilee Award, 2011; Padmasree Dr. I.V. Subba Rao Memorial Award from Rithunestam, 2012; International Potash Institute-FAI Award, 2012. Chief Editor, Indian Society of Dryland Agriculture, 2009 to 2013.

He is Fellow of Indian Society of Soil Science (FISSS); National Academy of Agricultural Sciences (FNAAS) and Indian Society of Pulses Research and Development (FISPRD).

Dr. AVR Kesava Rao is presently Scientist at ICRISAT, Hyderabad and has about 30 years of experience in teaching, research and extension in Agroclimatology. He has expertise in database management, agroclimatic characterization, climate change impact assessment, remote sensing and GIS techniques. He has practical experience in formulating agrometeorological advisories to the small farmers to improve their resilience to weather aberrations based on medium range weather forecasting in India. He is actively involved in capacity building of various stakeholders in climate change issues with particular reference to agriculture and use of crop simulation models in DSSAT.He developed software for agroclimatic analysis including water balance and estimating length of rainfed crop-growing period under different climate change scenarios. Undergone several professional trainings on various aspects of agrometeorology, remote sensing,GIS, climate science in India, Hadley Met Centre, UKand in crop modelling and precision farming at the University of Florida, USA.He extensively worked on GIS Spatial Analysis techniques for analyzing the spatial and temporal variations of soil moisture, rainfall and other climatic variables. He guided a few M.Sc. and Ph.D. students in Agroclimatology. Dr Kesava Rao has visited several countries viz., USA, UK, Kenya, South Africa, Thailand, Vietnam and Philippines.

 **Dr. Suhas Pralhad Wani** is currently Director of ICRISAT Development Center at ICRISAT, Patancheru, Hyderabad. His responsibilities include scaling-up center thru convergence and integration of technologies to achieve desired impact with a team of 27 scientists and 128 staff members. He also serves as Project Coordinator for Multi country Watershed Project in Asia and leading multidisciplinary team of scientists.

During the last three decades Dr.Wani has worked in the area of watershed management and water use efficiency (Green, blue and grey) in field to catchment scale. As a part of catchment scale, he studied, research use of degraded upstream lands rehabilitation using *Jatropha* and *Pongamia*, biofuel plantations for rehabilitation. His expertise include in implications of biofuel on water scenarios and food production, lvelihoods improvement for small farm holders, scaling-up and scaling-out technologies and trade offs in different sectors etc.

Dr. Wani's areas of specialization include, biofuels using non-edible oil seeds like *Jatropha* and *Pongamia*, their agronomy, water requirement and use for rehabilitation of degraded lands ensuring ecosystem services enhancement in the catchments, Integrated Watershed Management, Wasteland Development, Wastewater management , Biodiesel plantation, Integrated Nutrient Management, C Sequestration etc.

Dr. Wani has visited many countries on official missions and to undertake partnership research and discussions; participation in international conferences. that include Australia, Brazil, Canada, Cambodia, China, France, Germany, Kenya, Indonesia, Italy, Malaysia, Mexico, Myanmar, Nicaragua, Niger, Nigeria, Philippines, Singapore, Senegal, Switzerland Tanzania, Thailand, Syria, Ethiopia, Stockholm, USA, UK and Vietnam etc.

Dr. Wani published about 513 research papers in national and international journals, books and conference proceedings.

Dr. G. V. Ramanjaneyulu is currently Executive Director, Centre for Sustainable Agriculture, Hyderabad. He has studied B.Sc (Agric) at Bapatla and completed M.Sc (Agric Extn) from Andhra Pradesh Agricultural University, Hyderabad. He pursued his Ph.D in Agricultural Extension from Indian Agricultural Research Institute, New Delhi.

He was Selected for Indian Revenue Service in 1995 but preferred to work in agricultural research. He worked as Senior Scientist (Agrl. Extension) at Directorate of Oilseeds Research, Hyderabad for a decade. During that period he was Team leader for several projects studying the impacts of improved technologies on farming and farmers. He Course Director for several International and National Training programs on Oilseed production

In 2003 he resigned and working as Executive Director, Centre for Sustainable Agriculture, Secunderabad, India. He pioneered the concept of Non Pesticidal Management (NPM) in agriculture and experimented in Punukula, village in Khammam dist of Andhra Pradesh (AP) and made completely pesticide free without effecting the yields. He also established the first completely organic village Yenabavi in Warangal dist.

Dr. Pragati Pramanik Maity is currently working as Scientist at Center for Environment Science and Climate Resilient Agriculture, Indian Agricultue Research Institute (IARI), New Delhi. As part of Climate Change Challenge Programme at IARI, She has worked in assessment of changing trend of weather, soil and water parameters in fresh groundwater zones of Haryana, assessment of soil physical health and sustainability index of conservation and conventional agricultural system and evaluation of influence of tillage and residue incorporation on the dynamics of the N in the soil under wheat- maize rotation in semiarid conditions. She has received several awards and fellowships including ARS-NET by ICAR in 2008. Her specialization and research areas include agricultural physics, soil physics, greenhouse gases mitigation and ecosystem service.

Y.G. Prasad is a Principal Scientist (Entomology) at CRIDA, Hyderabad. He did his B.Sc (Ag.) from College of Agriculture, Bapatla, M.Sc and PhD in Entomology from IARI, New Delhi. He joined ICAR in 1992 at DOR and at CRIDA in 1999 as a senior scientist. He worked extensively on farmer participatory evaluation of bio-intensive IPM practices in three dryland crops, developed monitoring and evaluation strategy for sustainable initiative on cotton, identified and registered pest resistant castor germplasm accessions and developed granulosis virus as an eco-friendly biopesticide. He is trained in the field of pest forecast research and has led multi-institutional projects funded by the National Agricultural Technology Project and the National Agricultural Innovation Project and developed a crop pest decision support system for rice and cotton. He coordinated the country-wide efforts on preparation of district level agricultural contingency plans to meet aberrant weather situations. He is currently coordinating the Technology Demonstration Component of the National Innovations in Climate Resilient Agriculture (NICRA) project. He is a recipient of the NUFFIC International scholarship and is a member of many professional societies. He has to his credit 50 research publications, authored 31 books & technical bulletins, 11 book chapters and several popular articles. He is bestowed with the Bap Reddy Award for IPM by the Plant Protection Association of India.

Dr. J.V.N.S. Prasad is currently working as Principal Scientist (Agronomy) at ICAR Central Research Institute for Dryland Agriculture (CRIDA), Hyderabad. Earlier, he worked as Scientist at Indian Grassland and Fodder Research Institute (IGFRI), Jhansi. He has 16 years of research experience and handled several research and consultancy projects on various themes related to rainfed agriculture. His research interests include livelihood improvement of communities through productivity enhancement of crops, cropping systems and land use systems under rainfed conditions, adaptation and mitigation to climate change and climate resilient agriculture. He is instrumental in identification of K-636 variety of leucaena and paired row cultivation of eucalyptus and involved in scaling up of these technologies in association with paper industry. As a team member, evaluated developmental projects of the state government/ NABARD/ ministries such as AP Micro irrigation project, Community managed sustainable agriculture, Watershed management projects, assessment of the environmental services of MGNREGA funded by GIZ/Ministry of RD, Biennial Update Report of MOEF, etc. Dr. Prasad won S. Deshmukh Young Agronomist Award for the year 2001-2002 by the Indian Society of Agronomy. He has published more than 70 research papers and recipient of awards in the field of agronomy.

Dr. Himanshu Pathak is **Professor at** Center for Environment Science and Climate Resilient Agriculture, Indian Agricultue Research Institute (IARI), New Delhi and Principal Investigator, Climate Change Challenge Programme at IARI, New Delhi. He has worked as a visiting scientist in the University of Essex, United Kingdom; International Rice Research Institute (IRRI), Philippines; CSIRO Land and Water, Australia and Institute of Meteorology and Climate Research, Germany. During 2007-2009, he worked as the Co-Facilitator, Rice-Wheat Consortium (RWC), IRRI-India, New Delhi. He was the Lead Author, Inter-Governmental Panel on Climate Change (IPCC) 5th Assessment Report. Currently he is the Coordinator of DARE/ICAR for the climate negotiations in the United Nation Framework Convention on Climate Change.

Dr. Pathak is Fellow of the Indian National Science Academy (FNA), National Academy of Agricultural Sciences (FNAAS) and West Bengal Academy of Science and Technology (FWAST). He is the recipient of the Alexander von Humboldt Fellowship (FAvH) of Germany, Dr. BC Deb Memorial Award of the Indian Science Congress Association (ISCA), Golden Jubilee Commemoration Young Scientist Award of the Indian Society of Soil Science, BOYSCAST Fellowship of Department of Science and Technology, Govt. of India, Young Scientist Award and Prof. SK Mukherjee Commemoration Award of ISCA. Currently he is the Editor-in-Chief, Current Advances in Agricultural Sciences, Editor, Greenhouse gases: Science and Technology and Editor, Indian Journal of Agronomy.

Dr.Pathaks published more than 200 research papers and 10 books and has h-index of 44, i10-index of 100 with more than 6000 citations in international literature.

Dr. Rajeswar Jonnalagadda is a Senior Subject Specialist on Climate Change & Environment, at *Intercooperation Social Development India* (ICSD). His current responsibilities include Knowledge Management on Climate Change, Monitoring & Evaluation of Climate Change

Action-Plans in different sectors, and also Training & Capacity Building on *Mitigation&Adaptation* strategies for low-emission development, at national level and in the South Asia region. He pursues policy research towards integrating portfolios of Climate Change Adaptation (CCA), Disaster Risk Reduction (DRR) and Sustainable Development Goals (SDGs) into *Low-Emission Development and Inclusive Growth*. He is included in Roster of Experts for UNFCCC Adaptation Fund in Environmental and Social Risk Management

Prior to joining Intercooperation, he worked for more than five years as Director (Training & Research) of the Centre for Climate Change and Disaster Management at Dr.MCR Human Resource Development Institute, Hyderabad, India – a governmental administrative training institute. He was responsible for designing and developing training modules & learning materials on Mitigation and Adaptation measures against climate change. Earlier, for about twelve years, he worked as an Academic Editor & Consultant for UNESCO global project on Knowledge Management for Sustainable Development,and also as a Research Fellow in Abu Dhabi, UAE, on Integrated Approach to Water and Energy Systems.Prior to that, he did his Post-Doctoral research at Center for Ecological Sciences of the *Indian Institute of Science*, Bangalore, India. He pursued his M.Sc. and M.Phil. in Life Sciences from the University of Hyderabad and Ph.D.in Life Science from Jawaharlal Nehru University, Delhi, India

Dr. K. Ravi Shankar is a Senior Scientist (Agril. Extension) at the Central Research Institute for Dryland Agriculture (CRIDA), Hyderabad. He has 15 years working experience in drylands and his major research areas are Farmers' Perceptions, Attitudes and Adaptation measures tow ards Climate Change, Indigenous Rain Forecasting (ITKs) in Agriculture, Participatory Natural Resource Management in Drylands, Adoption studies and Information and Communication Technologies (ICTs). His Ph.D. thesis (on Indigenous Weather Forecasting) summary was entered as an International Compilation in Dissertation Initiative for the Advancement of Climate Change Research (DISCCRS) with Citation in 2006. He presented a paper on "Farmers' Knowledge Perceptions and Adaptation Measures towards Climate Change in South India" at Gold Coast, Queensland, Australia during June 29-1 July, 2010. He organized nearly 50 trainings at CRIDA on different topics viz., Integrated Watershed Management; Integrated Nutrient, Pest and Disease Management in

Drylands; Organic Farming in Rainfed Agriculture; Agricultural Mechanization in Drylands; Participatory Natural Resource Management for Enhancing Water Productivity in Drylands for middle level extension functionaries from state departments of agriculture/horticulture/animal husbandry all over India. He received the Young Scientist Award from Society of Extension Education, Agra at the 7th National Extension Education Congress on "Translational Research-Extension for Sustainable Small Farm Development" at ICAR RC for NEH Region at Barapani, Meghalaya during November 8-11, 2014. He has to his credit 13 research papers (two international), three books, six book chapters, 12 popular articles, three extension bulletins, two e-publications and 20 papers in proceedings of seminars/symposia/conferences. He is associate editor for International Journal of Climate Change Strategies and Management based in London.

Dr. R. Sudhakar Rao worked as Director of Research at Acharya N.G. Ranga Agricultural University (ANGRAU) and he has 35yrs of experience in teaching, research, extension and administration. He has published 61 original research papers in national and international journals of repute besides 14 papers presented at International seminars. As Director of Research he was instrumental in releasing 23 improved varieties in different crops namely rice, sorghum, finger millet, foxtail millet, pearl millets, redgram, bengalgram, soybean, groundnut, castor, sunflower and sugarcane which have been developed during his tenure as Director of Research at different research stations of the University. These varieties have been released to farmers during 2012. He secured several national and international awards for his contributions in agricultural research and development. He visited Kenya, Egypt, Israel, UK in connection with several research forums.

G. Kiran Kumar is currently working as Consultant (Agriculture) at Centre for Good Governance, Hyderabad. Prior to joining CGG in January 2013 after, he had five years of professional experience in corporate sector and was involved in Channel Sales, Technology Implementation and Marketing. He has a Bachelors degree in Agriculture and graduated from IIM Lucknow with a Post Graduate Diploma in Agri Business Management. Kiran Kumar was associated with Godrej Agrovet Ltd. and Rallis India Ltd. during his corporate stint.

At CGG, his work involves ideating farm centric interventions with respect to farm enterprise profitability and agri business activities in farming and allied enterprises, rural development. He is involved in conducting Impact Assessments and Evaluation Studies on assigned subjects, Improvement of Service Delivery of client departments. These are being achieved through discussions, adoption of best practices, on field surveys, key informant discussions. Preparation of concept notes, project proposals, data analysis, liaison with client departments are some of the routine tasks.

His key interests and expertise lie in community supported farm ventures, agriculture technology management and food & agri business industry.

Dr. K. Padmaja has 10 years of professional experience in Agricultural Extension sector in the Department of Agriculture, Government of Andhra Pradesh and 5 years of Agricultural Research experience in Directorate of Rice Research, an ICAR(Indian Council of Agricultural) Institute. Her education is primarily in the area of Agriculture with a specialization in Agronomy.In addition to that she did Post Graduate Diploma course in Agricultural Extension Management from MANAGE, Hyderabad. Presently she is working with Centre for Good Governance, Hyderabad on deputation from Dept. of Agrl, Telangana state. At CGG, she is involving in field surveys, impact assessments, action research, monitoring of the projects within the Agricultural and Rural Management Group of CGG and currently working with "Farm Viability Centric Interventions in Agriculture and Allied Sectors" project. She has published 10 research papers in the area of Agriculture in different reputed journals and has undergone trainings in the areas of Facilitator Development Skills, ICT in Agriculture, Revamping Delivery Systems, Ethics in Administration, Service Matters etc.